Studies in Optimization

Studies in Optimization

D. M. BURLEY

A HALSTED PRESS BOOK

JOHN WILEY & SONS
New York—Toronto

First published by
International Textbook Company Limited
Kingswood House, Heath & Reach, nr Leighton Buzzard, Beds.

First published 1974

Published in the U.S.A. and Canada by Halsted Press, a Division of John Wiley & Sons, Inc., New York.

Library of Congress Cataloging in Publication Data

Burley, David Michael.
Studies in optimization.

"A Halsted Press book."
1. Mathematical optimization. I. Title.
QA402.5.B89 515 74–8456
ISBN 0–470–12410–5

Printed in Great Britain by Galliard (Printers) Ltd,
Queen Anne's Road, Gt Yarmouth, Norfolk.

Acknowledgements

I would like to express my thanks to various people who have helped in the preparation of this book: Professor P. C. Kendall for first suggesting that the work be written in book form, Professor C. Storey of Loughborough University of Technology, Dr V. C. L. Hutson for reading, commenting and advising on the manuscript, and Mrs D. Wood for expertly typing the manuscript.

Contents

* Chapters and sections marked with an asterisk contain more difficult material. See paragraph two of Preface.

* Chapters and sections marked with an asterisk contain more difficult material. See paragraph two of Preface.

Preface

This book developed out of two courses of lectures that I have given on optimization. One course was aimed at second year undergraduates interested in the more practical side of the subject. This included work on hill climbing, dynamic programming and to a lesser extent variational methods. The other course was to final year students who were ultimately to take a course on optimal control theory. Consequently the emphasis was more on the Euler equation and its various aspects, with dynamic programming as a tool for solving such problems. The remainder of this book reflects my own interest in optimization which grew out of physical problems and the solution of differential equations.

The students attending the courses already mentioned and also final year engineering students who attended a similar but much abbreviated course had all studied a first course in calculus. They were familiar at least with elementary partial differentiation and vector calculus. A suitable scheme of study for a student with this minimum background would be chapters 1, 2, 3, 4, 5, and sections 6.1, 6.2, 7.1, 10.1, 10.2, 12.1, 12.2. The remaining chapters, 8, 9, 11, 13, and the remaining sections of the other chapters demand a much stronger mathematical background and could be omitted until the reader has acquired the necessary skill. For this reason the sections that I consider to be more difficult and that can be left out without significantly affecting the subsequent work are marked with an asterisk (*). They should all be studied eventually, however, since the more advanced mathematics provides a deeper understanding of optimization methods. It is inevitable that in a book which attempts a broad look at a subject there will be some differences in the required level of the mathematics. For this reason the division is made clear by the asterisks. It is hoped therefore that a variety of readers in mathematics, engineering or the physical sciences will be able to select a scheme of study for themselves, concentrating on their own particular interest. To assist the specialist reader considerable continuation reading is cited in the text. It will also help to counteract the impression given by many textbooks that the subject comes to an end at the last page; indeed it is my opinion that books should be open ended.

D. M. Burley
University of Sheffield

Chapter I

Classical Results

1.1 Introduction

The problem of optimizing assets has been an important consideration over many centuries. Before the advent of the calculus some problems could be solved (or had known solutions) but there were serious technical barriers to further progress. Once the new calculus had been established, however, most of the important results on necessary conditions for extrema were quickly established. Over the following two hundred and fifty years there was a slow but steady development of the subject; increasingly complex situations were considered, for example, problems with constraints and variational problems, and these were developed to high levels of sophistication. However, these classical results, carefully studied over the preceding centuries, foundered on one major difficulty, that these exact methods gave final numerical problems which were often quite intractable. With the advent of the electronic computer a new injection of life came into the subject. The computer provided just the means of resolving many of these numerical problems, but it was often found that these classical methods did not give the best computational methods. Consequently a set of *algorithms* for solving various classes of such problems developed. It is a mixture of this classical work and these computational algorithms that the book describes. The later chapters will be devoted largely to applications of the work in a variety of old and new contexts.

The main stimulus for much of this modern work came from the space industries. In such very expensive fields, even small savings of 1 or 2% can give enormous cash savings. Considerations of this sort led to considerable effort and energy being put into the related fields of optimization and control. In general, the large scale industries with expensive plant, chemical engineering, control engineering, steel, aircraft, have benefited most from this work. However, this is not the case in the best used of all the optimization methods described in the book, the linear programming contained in section 3.4. Here not only the large industries but quite modest enterprises have used to advantage the linear programming packages prepared by computer consultant firms.

The first major problem that occurs in any optimization problem is to set up a *cost function*. It frequently proves to be a most difficult job since there can be so many ways of doing it. For instance, in the operation of a factory a cost function could be the time spent to produce an item, or the total cost of running the factory, or the profitability of the factory, or, more frequently, a much more subjective value. Often the simple idea of which cost function to choose can lead to absurd, quite unacceptable results. For example, if the cost of running a factory is chosen as the cost function, the minimization would often lead to the answer 'close down the factory' when it will cost zero to run. It is clear that this is not what is required. The discussion of a suitable cost function usually leads to a careful analysis of the aims and objects of the situation under consideration. It becomes almost a philosophical discourse.

The cost function will depend on a set of parameters (or even a set of functions) and the object is, out of all allowable sets of parameters, to choose the set that provides the *optimum* cost. For instance, if the minimum of a cost function, which depends on several parameters, is required, then an optimum set of parameters is one for which the cost function is less than for all other sets.

Problems which come from real industrial situations are usually very complex. It is, however, desirable to illustrate them by the simple model problems that will be encountered in this book and which can be used as a platform for such further work. At this point various examples will be stated with comments, but without solutions. At a later stage most of these problems will be reconsidered and solved.

The first and best known method of obtaining maxima and minima is to solve the equation $dy/dx = 0$ (or its extension to several variables). This can be used, for instance, in the following:

Example 1 Find the maximum of the function

$$y = x^2 \exp(-x^2).$$

A more complicated example that can be approached by this traditional method is:

Example 2 Find the first minimum in the region $x \geq 0$ of the Bessel function defined by

$$J_0(x) = \frac{2}{\pi}\int_0^{\frac{1}{2}\pi} \cos(x \sin\theta)\, d\theta.$$

While dJ_0/dx can be calculated easily it should be noted that every function evaluation requires the computation of an integral. Since this is a very time-consuming operation it is absolutely essential that every piece of information should be used as efficiently as possible. By considering even more difficult functions it is often found that the differentiation of the function itself is quite impossible. A simple physical example illustrates this point to some extent.

Example 3 Find the angle of projection α which gives the maximum horizontal range of a projectile fired with speed v in a medium whose resistance per unit mass is $-k\mathbf{v}$.

The equations of motion

$$\ddot{\mathbf{r}} = -g\hat{\mathbf{j}} + k\dot{\mathbf{r}}$$

can be solved to give the range $R(\alpha)$ from

$$R(\alpha) = \frac{v}{k}(1 - e^{-T}) \cos \alpha,$$

where T is a solution of

$$T = \left(\frac{kv}{g} \sin \alpha + 1\right)(1 - e^{-T}). \tag{1.1}$$

While the maximum range is obtained from solving $dR/d\alpha = 0$, it should be noted that T is a function of α, defined implicitly as a solution of (1.1). Hence the evaluation of $dR/d\alpha$ is a fairly complicated procedure, but it can be done. Generalizing this example to having resistance proportional to the square of the velocity would mean that the differential equations themselves could not be solved explicitly and the differentiation would then be out of the question.

In this example the range R is maximized subject to the *constraint* (1.1), which can be interpreted physically as giving the (non-dimensional) time required for the projectile to reach ground level. Constraints play a very important part in many optimization problems and in many modern problems they are of prime interest. The following is a more straightforward geometrical problem involving a constraint.

Example 4 Find the maximum of the function $f(x, y) = x + y$ subject to $x^2 + y^2 = 1$.

The problem asks for the maximum of f but the only points that are to be considered lie on the unit circle. Lagrange showed how to tackle such problems with the use of undetermined multipliers.

Of late this idea has been developed considerably to cope with constraint problems in control theory. These problems usually involve *inequality* constraints illustrated by the post-office problem.

Example 5 To go into a post-office sack, a rectangular parcel of sides x_1, x_2, x_3 must satisfy the constraints $x_i \leq 42$ inches, $i = 1, 2, 3$ and $x_1 + 2x_2 + 2x_3 \leq 72$ inches. Find the maximum volume, $V = x_1 x_2 x_3$ subject to these constraints.

This is a non-trivial problem involving inequality constraints. Other constraints are hidden in this problem since all the sides must have positive length and hence $x_i \geq 0$, $i = 1, 2, 3$ must be added to the above. In many modern problems, it is just this kind of constraint that is operative.

A completely different class of problems occurs where the desired optimum is not a number or set of numbers, but a function. This function is chosen from a given class of functions to optimize a cost which is usually in the form of an integral. This general type is usually called a *variational problem*. A particular example to illustrate such work concerns the shortest distance or geodesic problem.

Example 6 Find the continuous curve joining two given points in a plane, which has minimum length.

Thus out of all continuous curves that join the two points choose the one that minimizes the length. It is a reasonably straightforward exercise to set up the necessary condition for the minimum of this variational problem and the resulting differential equations can be solved easily. For slightly more complex problems these equations are quite impossible to solve explicitly and they must be treated numerically. It is often easier to tackle the original problem by a direct numerical technique, and both methods of attack will be described in later chapters. Many of these variational problems and their solutions were well known to the Greeks, but the methods of proof were not established until after the 16th century. The following is an example.

Example 7 Find the plane closed curve of given length that maximizes the enclosed area.

When land was conquered by the Greeks it was divided up by staking out areas of land with equal perimeters. It is well documented that the astute warriors chose the areas most nearly circular.

Once more it should be noted that this problem is a constrained one since the curve must have given length. Such constraints are very common in variational problems and a more modern example shows this.

Example 8 An aeroplane has weight W, thrust T, drag D as known constants. Its height h, angle of flight θ, and velocity v satisfy

$$\frac{dh}{dt} = \frac{v}{W} \frac{(T - D)}{1 + (v/g)(dv/dh)} = v \sin \theta.$$

For a given $\theta(t)$ these differential equations can be solved for $v(t)$ and $h(t)$. Find the flight path (i.e. $\theta(t)$) which gives the minimum time for the aircraft to climb from horizontal flight with velocity v_1 at height h_1 to horizontal flight with velocity v_2 at height h_2.

Similar problems occur in artificial satellite trajectories but in these problems the differential equations and constraints are impossible to write down in simple form and often occur as numbers inside a computer. The extreme complexity of the mathematics of necessity implies that numerical procedures must be used. Out of the many methods developed to cope with such problems one particularly important method to emerge is that of *dynamic programming*. It deals with a multistage decision process and a typical discrete problem for solution is as follows.

Example 9 Using only the existing road system find the path that minimizes the distance between London's King's Cross and Charing Cross Stations.

At each road junction encountered on the journey a decision has to be made: whether to proceed left, right or straight on. It can be seen that provided the map is a restricted one and reasonable rules are obeyed there are only a finite (but extremely large) number of possible paths. These can be enumerated, their lengths computed and compared. The method of dynamic programming attacks just such a problem and can reduce the amount of work dramatically. In a well quoted example, which will be considered in a later chapter, the number of additions can be reduced from about one million to about two hundred.

Finally a simple but not untypical example illustrates problems in control theory.

Example 10 Find the forcing function $F(t)$ required to minimize the time taken to move a body at rest at $x = a$ to rest at the origin if

$$\ddot{x} = F \quad \text{and} \quad |F| \leq 1.$$

Two points should be noted. Firstly, it is a variational problem since a function $F(t)$ has to be selected and, secondly, it involves inequality constraints.

1.2 Standard Results

1.2.1 Weierstrass Theorem

An important idea that will be used throughout this book is that of *continuity*; it will be assumed that the reader has at least a good intuitive understanding of the idea. The Weierstrass theorem deals with an important application of continuity to optimization.

Theorem If a function, f, is continuous in a closed, bounded domain, D, then it attains its maximum and minimum values in D.

Proof First note that D is closed and hence contains its boundary. Secondly, it is necessary only to deal with the case of the maximum and the minimum follows similarly.

To be continuous in D the function cannot be infinite and hence there exist numbers K, L with the property $K < f < L$ for all points in D. Indeed there will be infinitely many of these upper bounds L so choose the least of them, say M.

Now argue by contradiction. Suppose f does not attain its maximum in D then $f < M$ for all $x \in D$. Since $(M - f) > 0$ for all $x \in D$ and since f is continuous then $(M - f)^{-1}$ is a continuous positive function in D. By the above argument it has an upper bound, U, and $0 < (M - f)^{-1} < U$ for all $x \in D$, or

$$f < M - \frac{1}{U}.$$

A contradiction has now been produced since $(M - 1/U)$ is an upper bound of f less than the *least* upper bound. It now follows that the maximum M is attained (similarly the minimum) and the theorem is proved.

For a single variable the theorem can be illustrated and a typical case is shown in figure 1.1. It can be seen that the function attains its maximum at X and minimum at T. The importance of the theorem is that provided continuity of the function is guaranteed in a closed domain then it is known that the maximum and minimum exist in or on the boundary of the domain. It is then worth while to set up machinery to look for the optimum value. The

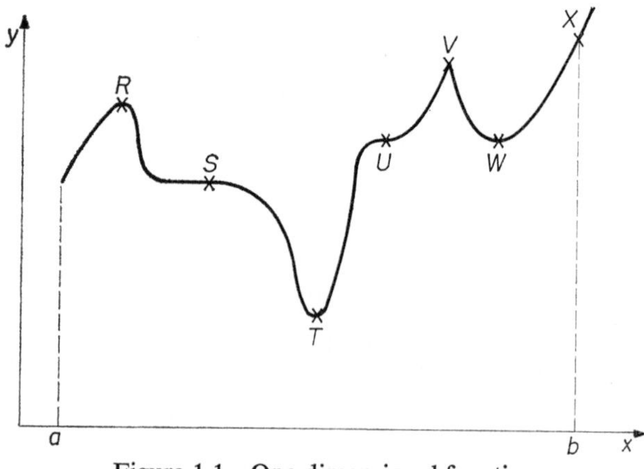

Figure 1.1 One dimensional function.

theorem, however, does not give any idea how to construct this machinery, nor does it give any information about whether the *global extremum* of the theorem is also a *local extremum*. A *global maximum* is defined just as in the theorem as the least upper bound of f for all $x \in D$. A *local maximum* on the other hand exists at a point x_0 if $f(x) \leq f(x_0)$ for *all points in the neighbourhood of* x_0. (Local and global minima are defined similarly.) Looking at figure 1.1 it can be seen that R, S and V are local maxima, T and W local minima, X is the global maximum and T is the global minimum. At the point S, $f(S) \geq f$ for all neighbouring points, but in fact $f = f(S)$ for all neighbouring points $x \leq x_S$. Such points are often identified separately as *weak* maxima as opposed to points like R which are called *strong* maxima. Finally it should be noted that if a global extremum is not a local extremum then it occurs on the boundary (e.g. X).

1.2.2 LOCAL MAXIMUM AND MINIMUM

Consider functions of a single variable which have continuous second derivatives. If such a function f has a maximum at $x = a$ then (for *all* sufficiently small h)

$$f(a + h) \leq f(a). \tag{1.2}$$

Taylor's theorem now gives

$$f(a + h) = f(a) + \frac{h}{1!}f'(a) + \frac{h^2}{2!}f''(a + \theta h), \qquad 0 < \theta < 1$$

or

$$f(a + h) - f(a) = hf'(a) + \tfrac{1}{2}h^2 f''(a + \theta h). \tag{1.3}$$

Now (1.2) states that the (LHS of (1.3)) ≤ 0 for all sufficiently small h; but the right-hand side of (1.3) is dominated by the term $hf'(a)$ which changes sign as h changes sign if $f'(a) \neq 0$. Hence a necessary condition for a maximum is that $f'(a) = 0$. If also $f''(a) < 0$, since f'' is continuous then $f''(x) < 0$ for $(a - k) < x < (a + k)$. By choosing $|h| < k$ equation (1.3) now says that equation (1.2) is satisfied if $f'(a) = 0$, $f''(a) < 0$. The corresponding result holds for a minimum.

RESULT I

A necessary condition for a local maximum or minimum of a function f at $x = a$ is that

$$f'(a) = 0.$$

If in addition

$$\begin{Bmatrix} f''(a) < 0 \\ f''(a) > 0 \end{Bmatrix}$$

then f has a local

$$\left.\begin{matrix} \textit{maximum} \\ \\ \textit{minimum} \end{matrix}\right\} \textit{at } x = a.$$

Geometrically this result has the usual interpretation that $f'(a) = 0$ implies a horizontal tangent and $f''(a)$ determines the concavity of the function. Examination of $f''(a)$ shows that if $f''(a) > 0$ then the curve is concave up while if $f''(a) < 0$ the curve is concave down. The difficult case for which no result has been obtained is the situation when $f'(a) = f''(a) = 0$. It is necessary in this case to look at further evidence, usually higher derivatives.

Example 11 (cf. example 1) Find the local and global maxima and minima of the function

$$f = x^2 \exp(-x^2) \quad \text{in} \quad 0 \le x \le \tfrac{1}{2}.$$

For local extrema

$$0 = f'(x) = 2 \exp(-x^2)(x - x^3)$$

$$f''(x) = 2 \exp(-x^2)(1 - 5x^2 + 2x^4).$$

At $x = 0$, $f'(0) = 0$, $f''(0) > 0$ and hence a local minimum exists at (0, 0). The other roots $x = \pm 1$ are outside the region of interest. The global extrema are not so easily obtained, but a sketch of the function reveals that global maximum is at $x = \frac{1}{2}$ and the global minimum is also the local minimum $f = 0$ at $x = 0$.

Example 12 Find the local extrema of

$$\text{(i)} \quad f = x^3 \qquad \text{(ii)} \quad f = x^4.$$

In both cases $f(0) = f'(0) = f''(0) = 0$ and nothing can be deduced from Result I. In the case (i) the function is always increasing with x and has a horizontal tangent at $x = 0$. This is the usual *horizontal point of inflexion* or *saddle point* illustrated in figure 1.1 at point U. On the other hand in the case (ii) $f \ge 0$ for all x and hence $x = 0$ is a local and global minimum of the function.

This example illustrates well the point that if $f'(a) = f''(a) = 0$ then either a maximum or a minimum or a saddle point can be obtained. These saddle points are comparatively rare in functions of a single variable but for functions of several variables they are very common.

Consider a function of two variables. The method of deducing the necessary conditions follow closely the method used above where it will have been noted that an analytical rather than geometrical approach was used. This choice was made since geometrical intuition becomes increasingly difficult to maintain as the number of dimensions rises while the analytical method demands no such intuition. If the chosen function $f(x, y)$ has a local maximum at (a, b) then (for all sufficiently small h, k)

$$f(a + h, b + k) \leq f(a, b).$$

Using Taylor's theorem again

$$f(a + h, b + k) - f(a, b) = (hf_x + kf_y) + \tfrac{1}{2}(h^2 f_{xx} + 2hkf_{xy} + k^2 f_{yy}) + \cdots \tag{1.4}$$

where all the derivatives in the right-hand side are evaluated at (a, b). For a maximum the (LHS of (1.4)) ≤ 0 for all small h, k but since h, k can both change sign, the dominant term in the RHS of (1.4), $(hf_x + kf_y)$, can also change sign. Thus a necessary condition for a maximum at (a, b) is $f_x = f_y = 0$ at this point. While this condition is necessary, it is not sufficient since it is now required that the second bracket in (1.4) is negative for all small h, k. Writing

$$h^2 f_{xx} + 2hkf_{xy} + k^2 f_{yy} = h^2 f_{xx}\left[\left(1 + \frac{kf_{xy}}{hf_{xx}}\right)^2 + \left(\frac{k}{h}\right)^2 \frac{(f_{xx}f_{yy} - f_{xy}^2)}{f_{xx}^2}\right]$$

it may be noted that if

$$(f_{xx}f_{yy} - f_{xy}^2) > 0$$

then the square bracket is always positive and if

$$f_{xx} < 0$$

then the right-hand side is always negative. Hence these conditions are sufficient for a maximum. If on the other hand $(f_{xx}f_{yy} - f_{xy}^2) < 0$ then second bracket of (1.4) can take both signs as h and k vary and a saddle point is obtained. Figure 1.2 illustrates a maximum, minimum and saddle point for a function of two variables.

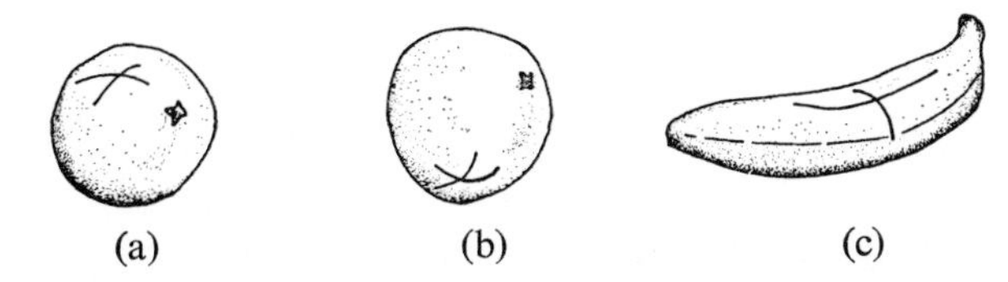

Figure 1.2 Two dimensional extrema: (a) maximum, (b) minimum, (c) saddle point.

RESULT II

A necessary condition for a local extremum of a function f at $(a_1, a_2, \ldots, a_n)$ is that

$$\frac{\partial f}{\partial x_1} = \frac{\partial f}{\partial x_2} = \frac{\partial f}{\partial x_3} = \cdots = \frac{\partial f}{\partial x_n} = 0 \quad at \quad (a_1, a_2, \ldots, a_n).$$

For a two variable function if in addition

$$J = \begin{vmatrix} f_{x_1x_1} & f_{x_1x_2} \\ f_{x_2x_1} & f_{x_2x_2} \end{vmatrix} > 0 \quad \textit{and} \quad \begin{Bmatrix} f_{x_1x_1} < 0 \\ f_{x_1x_1} > 0 \end{Bmatrix}$$

then the function has a local

$$\begin{Bmatrix} \textit{maximum} \\ \textit{minimum} \end{Bmatrix}.$$

It will be found that the *Jacobian* or *Hessian* J occurs on many occasions throughout this sort of work. Again if $J = 0$ or $f_{x_1x_1} = 0$ then Result II gives insufficient evidence to determine the type of extremum. For higher dimensions the corresponding results to ensure a maximum or minimum can be found in Mirsky (1955) or in problem 6.

Example 13 Show that $f = 3xy - x^3 - y^3$ has a local maximum at (1, 1).

Now $f_x = 3y - 3x^2 = 0$, $f_y = 3x - 3y^2 = 0$ or $y = x^2$ and $x = y^2$ giving

$$0 = y^4 - y = y(y - 1)(y^2 + y + 1).$$

Possible local extrema occur at (0, 0) and (1, 1). Calculating

$$f_{xx} = -6x, \qquad f_{xy} = 3, \qquad f_{yy} = -6y.$$

At (0, 0), $f_x = f_y = 0$, $J = -9$ and hence a saddle point. At (1, 1), $f_x = f_y = 0$, $J = 27$, $f_{xx} = -6$ and hence a maximum.

1.2.3 CONSTRAINTS

It was indicated in section 1.1 that constraints play an important part in optimization and it will now be shown how Lagrange successfully attacked problems involving equality constraints. Consider first a straightforward approach to example 4.

Example 4 Find the maximum of the function $f(x, y) = x + y$ subject to $x^2 + y^2 = 1$.

Method 1 Now $y = \pm(1 - x^2)^{\frac{1}{2}}$ on the unit circle and hence with the constraint taken care of

$$f = x \pm (1 - x^2)^{\frac{1}{2}}.$$

This is now a single variable problem and Result I can be used. It is found that $(2^{-\frac{1}{2}}, 2^{-\frac{1}{2}})$, $(-2^{-\frac{1}{2}}, -2^{-\frac{1}{2}})$ give $df/dx = 0$ and these correspond to a maximum and minimum respectively.

Method 2 An alternative method is to parameterize the constraint curve. In this case $x = \sin\theta$, $y = \cos\theta$ is a suitable parametric form and the function f becomes

$$f = \sin\theta + \cos\theta,$$

a function of a single variable. Result I reproduces the same answers as method 1.

Both these methods are valuable but for more difficult constraints, although the same procedure is possible theoretically, they are usually difficult to implement. The general problem for two variables is:

General Problem (for two variables). Find the local extrema of $F(x, y)$ subject to the constraint $g(x, y) = 0$.

Assume that (a, b) is a local maximum; then for all sufficiently small h, k

$$F(a + h, b + k) \leq F(a, b)$$

and since points must lie on the constraint

$$g(a, b) = g(a + h, b + k) = 0.$$

Applying the usual argument, the Taylor expansions give

$$0 \geq F(a + h, b + k) - F(a, b) = (hF_x + kF_y) + 0(h^2) \tag{1.5}$$

and

$$0 = g(a + h, b + k) - g(a, b) = (hg_x + kg_y) + 0(h^2). \tag{1.6}$$

Now (1.6) shows that h and k are not independent. Let one of these, h say, be chosen as independent variable and k is now determined. Combining (1.5) and (1.6)

$$0 \geq h(F_x + \lambda g_x) + k(F_y + \lambda g_y) + \cdots,$$

and choose λ so that $(F_y + \lambda g_y) = 0$. Since h is the independent variable it can be chosen to have either sign and this inequality cannot be satisfied unless $(F_x + \lambda g_x) = 0$ also. The basic constraint equation $g(x, y) = 0$ must also be satisfied and these three equations therefore give a necessary condition for a maximum.

This general method can be used for problems involving more variables and more constraints, it is called *Lagrange's method of undetermined multipliers*. The basic result for this method is as follows.

RESULT III

A necessary condition for a local extremum of the function $F(x_1, x_2, \ldots, x_n)$ *subject to* $f_i(x_1, x_2, \ldots, x_n) = 0$, $i = 1, 2, \ldots, m$, $(m < n)$ *is given by*

$$\frac{\partial F}{\partial x_1} + \lambda_1 \frac{\partial f_1}{\partial x_1} + \cdots + \lambda_m \frac{\partial f_m}{\partial x_1} = 0$$

$$\frac{\partial F}{\partial x_2} + \lambda_1 \frac{\partial f_1}{\partial x_2} + \cdots + \lambda_m \frac{\partial f_m}{\partial x_2} = 0$$

$$\vdots$$

$$\frac{\partial F}{\partial x_n} + \lambda_1 \frac{\partial f_1}{\partial x_n} + \cdots + \lambda_m \frac{\partial f_m}{\partial x_n} = 0 \tag{1.7}$$

together with the constraint equations satisfied.

An equivalent formulation of Result III is to find the necessary condition for an extremum of the unconstrained function

$$F^* = F + \lambda_1 f_1 + \cdots + \lambda_m f_m.$$

Result II gives the conditions (1.7) which together with the constraint equations gives Result III. This method of working usually turns out to be the most convenient in practice.

Example 14 Find the local extrema of the function

$$F = x + y \quad \text{subject to } x^2y^3 + y^2x^3 = 64.$$

Construct

$$F^* = x + y + \lambda(x^2y^3 + y^2x^3 - 64)$$

and calculate

$$0 = \frac{\partial F^*}{\partial x} = 1 + \lambda(2xy^3 + 3y^2x^2)$$

$$0 = \frac{\partial F^*}{\partial y} = 1 + \lambda(3x^2y^2 + 2yx^3),$$

which must be solved together with

$$x^2y^3 + y^2x^3 = 64.$$

Subtracting the first two of these equations gives $0 = \lambda 2xy(y^2 - x^2)$. Of all the possibilities in this equation the only one to satisfy the constraint equation is $x = y$ and the only real solution is $x = y = 2$, $\lambda = -1/80$, $F = 4$. Thus there is only one possible candidate for an extremum, but further evidence is required before its type can be decided. It is left to the reader to show that the point is in fact a minimum.

One of the strengths of the Lagrange method is that even when the λ has been introduced it is not required for the final solution and hence need not be evaluated explicitly. It is, however, a surprising fact that often in physical and geometrical problems the Lagrange multipliers themselves have a useful interpretation and are frequently required in the course of the calculation. This fact will be seen in later chapters but a simple example serves as an illustration.

Example 15 A point $P(x, y)$ lies on the curve $x^2 - xy + y^2 = 1$. It is required to find the minimum distance from P to the origin as P varies along the curve.

The problem is to minimize $f = x^2 + y^2$ subject to $x^2 - xy + y^2 = 1$. Construct

$$f^* = x^2 + y^2 - \lambda(x^2 - xy + y^2 - 1)$$

and calculate

$$0 = \frac{\partial f^*}{\partial x} = 2x - \lambda(2x - y)$$

$$0 = \frac{\partial f^*}{\partial y} = 2y - \lambda(-x + 2y).$$

At the extremum

$$f = f^* = f^* - \tfrac{1}{2}x\frac{\partial f^*}{\partial x} - \tfrac{1}{2}y\frac{\partial f^*}{\partial y} = \lambda,$$

and hence the square of the required distance is given by the multiplier λ. Solving the three equations gives $\lambda = 2$ or $\frac{2}{3}$ and choosing $\lambda = \frac{2}{3}$ (or distance $= (\frac{2}{3})^{\frac{1}{2}}$) produces the minimum. In this case the distance can be obtained without ever explicitly calculating x and y. In larger calculations this type of result can be useful and save a large amount of computation.

1.3 Problems to be Tackled

Having established three important results it is now possible to solve some of the more straightforward examples in section 1.1 or at least to understand the technical problems involved.

Example 2 Find the first local minimum for $x > 0$ of the Bessel function defined by

$$J_0(x) = \frac{2}{\pi}\int_0^{\frac{1}{2}\pi} \cos(x \sin\theta)\, d\theta.$$

Using Result I it is necessary to solve

$$\frac{dJ_0}{dx} = -\frac{2}{\pi}\int_0^{\frac{1}{2}\pi} \sin\theta \sin(x\sin\theta)\, d\theta = 0.$$

It is clear that this equation is a difficult one and will take a considerable numerical effort to extract the value of x which makes the integral zero. (In fact readers who are familiar with Bessel functions will know that $J_0(x)$ and $J'_0(x) = -J_1(x)$ are well tabulated functions and this problem is not quite as fierce as it looks. This is because of the importance of Bessel functions in mathematical physics.) The methods that try numerically to obtain the minimum in the smallest number of evaluations of J_0 and J'_0 come under the general heading of hill climbing techniques and will be discussed in the next chapter.

The following is a comparable several variable example.

Example 16 An elastic string, with modulus λ and natural length a, when extended a distance l has potential energy $V = \frac{1}{2}\lambda l^2/a$. Three elastic strings OP, BP, CP, each of natural length $\frac{1}{2}a$ and modulus λ, 2λ, 3λ respectively, are fixed at O, B, C and tied together at P. Relative to axes Oxy the points O, B, C, P have coordinates (0, 0), (0, a), ($2a$, a), (x, y). Find the equilibrium point P by minimizing the total potential energy.

Now $OP^2 = x^2 + y^2$, $BP^2 = x^2 + (y - a)^2$ and $CP^2 = (x - 2a)^2 + (y - a)^2$ and the total potential energy V is

$$V = \left(\frac{\lambda}{a}\right)[(OP - \tfrac{1}{2}a)^2 + 2(BP - \tfrac{1}{2}a)^2 + 3(CP - \tfrac{1}{2}a)^2].$$

To find the minimum, which is clearly a local minimum,

$$0 = \frac{\partial V}{\partial x} = \frac{\lambda}{a}\left(2(OP - \tfrac{1}{2}a)\frac{x}{OP} + 4(BP - \tfrac{1}{2}a)\frac{x}{BP} + 6(CP - \tfrac{1}{2}a)\frac{(x - 2a)}{CP}\right)$$

$$0 = \frac{\partial V}{\partial y} = \frac{\lambda}{a}\left(2(OP - \tfrac{1}{2}a)\frac{y}{OP} + 4(BP - \tfrac{1}{2}a)\frac{(y - a)}{BP} + 6(CP - \tfrac{1}{2}a)\frac{(y - a)}{CP}\right).$$

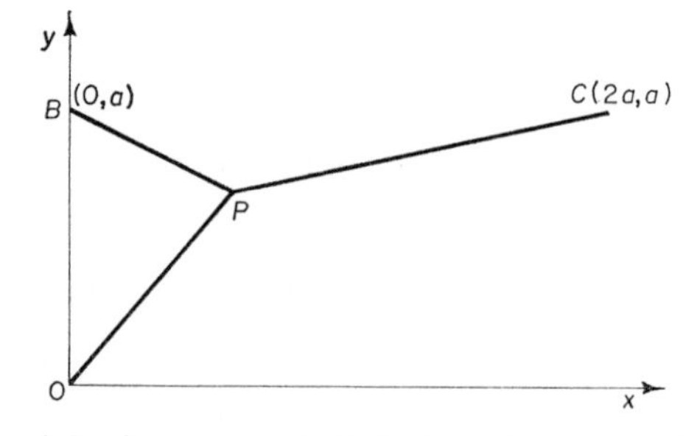

Figure 1.3 Arrangement of elastic strings, example 16.

Here again, while it is theoretically possible to solve these two equations for x and y, the hill climbing techniques of the next chapter help considerably in the practical calculation.

In the last two examples it was the local minimum that was sought, geometrical and physical considerations making this quite clear. In practical, non-scientific problems, however, it is the global result that is frequently required. These results are more difficult to obtain since they can occur in the interior of the region if the global extremum is also a local one or on the boundary if not. The methods that seek global extrema must therefore be equally efficient both in the interior and on the boundary.

The constraints in example 16 (that the strings do not become slack) need not be used provided reasonable initial estimates are taken. In other problems, however, the constraints are an essential part of the calculation. When such examples are studied it is possible to convert to an unconstrained problem by the use of Lagrange multipliers. However, their use can often lead to a problem more difficult than the original. This dilemma is illustrated below.

Example 3 Find the angle of projection which gives maximum horizontal range of a projectile fired with speed v in a medium with resistance proportional to the velocity.

It was asserted that the solution was given by

$$R(\alpha) = \left(\frac{v}{k}\right)(1 - e^{-T}) \cos \alpha$$

with

$$T = \left[\left(\frac{kv}{g}\right) \sin \alpha + 1\right](1 - e^{-T}).$$

Treating this problem with a Lagrange multiplier

$$R^* = \left(\frac{v}{k}\right)(1 - e^{-T}) \cos \alpha + \lambda \left\{\left[\left(\frac{kv}{g}\right) \sin \alpha + 1\right](1 - e^{-T}) - T\right\}$$

gives the equations

$$0 = \frac{\partial R^*}{\partial \alpha} = -\left(\frac{v}{k}\right)(1 - e^{-T}) \sin \alpha + \lambda \left(\frac{kv}{g}\right) \cos \alpha(1 - e^{-T})$$

$$0 = \frac{\partial R^*}{\partial T} = \left(\frac{k}{v}\right) e^{-T} \cos \alpha + \lambda \left\{\left[\left(\frac{kv}{g}\right) \sin \alpha + 1\right] e^{-T} - 1\right\}$$

together with the constraint equation. It now becomes a problem of whether to solve these three equations directly for α, T, λ or to tackle the original one directly. In this case enough progress can be made to take the former course of action but chapter 3 will look at the other course, hill climbing methods with constraints.

Finally look at the simplest of the variational problems to find out exactly what is involved.

Example 6 Find the continuous curve joining two given points in a plane and having minimum length.

Let the points be O (0, 0) and A (a, 0). Assume that the curve $y = y(x)$ is traversed from O to A so that x is always increasing. (There is no loss of generality since if the converse holds a shorter path can be constructed.) The length of this curve is obtained from $ds^2 = dx^2 + dy^2$ and gives

$$L = \int_0^a (1 + y'^2)^{\frac{1}{2}}\, dx.$$

Since the curve must pass through O and A then $y(0) = 0$ and $y(a) = 0$. Hence the problem consists of looking at each differentiable function of x satisfying the two end conditions, evaluating the integral and selecting the

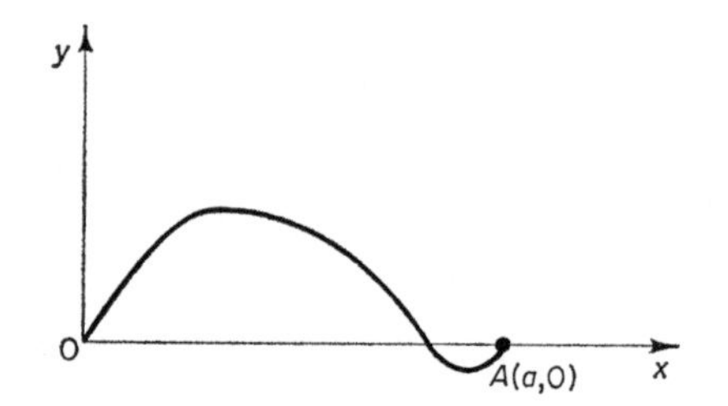

Figure 1.4 Test function for shortest distance, example 6.

function(s) which makes L a minimum. The obvious straight line answer is easy to justify for this simple case, but a general method of attack is required for this type of problem. In later chapters various aspects of variational problems will be considered and it will be seen how to apply such methods to physical and geometrical problems. The rather trickier control problems, example 10, will be considered when a little more mathematical apparatus has been established.

PROBLEMS

1. Find the local maxima and minima of the function

$$y = \frac{3x - 5}{x^2 - 1}.$$

Find also the global extrema in the region $x \geq 2$.

2. A ladder is to be taken horizontally round a right-angle bend in a corridor with branches of width a and b. Find the longest ladder that can be used.
3. If $f'(a) = f''(a) = 0$ and $f'''(a) > 0$ sketch the curve as it passes through $x = a$.
4. Show that

$$f(x, y) = x^4 + y^2 - 4xy^2$$

has only one local extremum.
5. Show that the sum of the squares of the distances of a point from the vertices of a triangle is a minimum when the point is at the centroid of the triangle.
6. Prove Result II for the three variable function $f(x_1, x_2, x_3)$.
7. Find the local maximum of the function

$$f = x^3 + y^3 + z^3$$

subject to $x^2 + y^2 - z^2 = 1$.

Chapter II

Hill Climbing without Constraints

2.1 Introduction

The present chapter is concerned with problems where the function to be optimized is a complicated one; examples 2, 3, 16 of chapter 1 illustrate this. A further example shows a problem where the function evaluation is a formidable task and the derivative calculation almost impossible.

Example 17 A pendulum consists of a rod of length $2a$ and mass M and it can swing about one end. A regulator consists of a mass m attached at a distance x from the pivot. Find the period of (large) oscillations of the pendulum if it is given an angular velocity ω while hanging in equilibrium. Has this period a local maximum or minimum as x varies?

If the rod makes an angle θ with the vertical then the equation of motion is

$$I\ddot{\theta} + g(Ma + mx)\sin\theta = 0$$

where $I = \frac{4}{3}Ma^2 + mx^2$. After two integrations the period $t(x)$ is given by

$$t(x) = 4\int_0^{\alpha} \frac{d\theta}{(\omega^2 - 4k^2\sin^2\frac{1}{2}\theta)^{\frac{1}{2}}},$$

where $k^2 = g(Ma + mx)/(\frac{4}{3}Ma^2 + mx^2)$ and $\alpha = 2\sin^{-1}(\omega/2k)$.

While it is theoretically possible to evaluate dt/dx the effort involved is large even if full use of elliptic integrals is made. Some indication of the type of behaviour expected can be obtained by looking at the small oscillation limit when $t(x) = 2\pi/k = A[(\frac{4}{3}C + y^2)/(C + y)]^{\frac{1}{2}}$, where A is a constant, $C = M/m$ and $y = x/a$. This function has a negative gradient at $y = 0$ and tends to $+\infty$ as $y \to +\infty$ and hence has a local minimum somewhere

between. It may be noted that such a minimum may be out of the range of interest since the regulator is fixed to the rod so that $0 \leq x \leq 2a$.

In other examples, example 2 for instance, the derivative is easy to obtain but both function and derivative evaluations are major tasks. Similarly in more than one dimension, example 16 for instance, the same sort of problems remain. The object, therefore, is to develop a series of numerical algorithms which give a reasonable assurance that the extrema can be obtained. It is clear that if an analytic solution can be obtained then this is preferable since it gives a much clearer mathematical or physical picture of the problem under study. If, however, the problem has to be studied numerically then it is essential that the work be done efficiently.

The present chapter is not exhaustive by any means (the important method of conjugate gradients is omitted completely) but is meant to give an indication of the main methods used. Indeed there are almost as many methods as there are workers in the field. Comprehensive reviews and bibliographies can be found in Beveridge and Schechter (1970), Murray (1972). One method, for instance, which will not be looked at in detail, involves a random search. Suppose it is required to maximize a given function. A point is chosen at random, the function evaluated there and this process is repeated as many times as possible. If the evaluation at the current point gives the largest function value yet calculated then it is retained as the best point, otherwise the next random point is chosen. The method has the considerable advantages of simplicity and of requiring little storage; it is, however, comparatively inefficient since it discards all the previously obtained information. Even though the amount of work increases exponentially with the number of independent variables, when this number is very large it is often the only one possible.

2.2 Single Variable Problems

2.2.1 Bracketing

It is most important as a first step in a calculation to get a rough idea of where to look for an extremum; a useful idea is to find two values that bracket the extremum. Suppose the maximum of a function $f(x)$ is sought and it is known that the maximum is in the region $x \geq a$. Choose an increment h and evaluate f at points $x_1 = a$, $x_2 = x_1 + h$, $x_3 = x_2 + 2h$, $x_4 = x_3 + 4h$, $x_5 = x_4 + 8h, \ldots$, that is doubling the increment at each stage. The evaluation is stopped if either the maximum is bracketed or if $x_i > X$, where X is a suitable large constant chosen at the start of the calculation. A maximum is bracketed if at some stage $f(x_i) > f(x_{i-1})$ *and* $f(x_i) > f(x_{i+1})$; the bracket is (x_{i-1}, x_{i+1}). If the *gradient* is known also

then if f' at x_j and f' at x_{j+1} are positive and negative respectively then the bracket for the maximum is (x_j, x_{j+1}). If the value of f increases until X is reached it is usually assumed that the function increases indefinitely; if on the other hand the value of X is reached and the function is still decreasing then $f(a)$ is usually taken as the maximum value. Figure 2.1 illustrates this procedure.

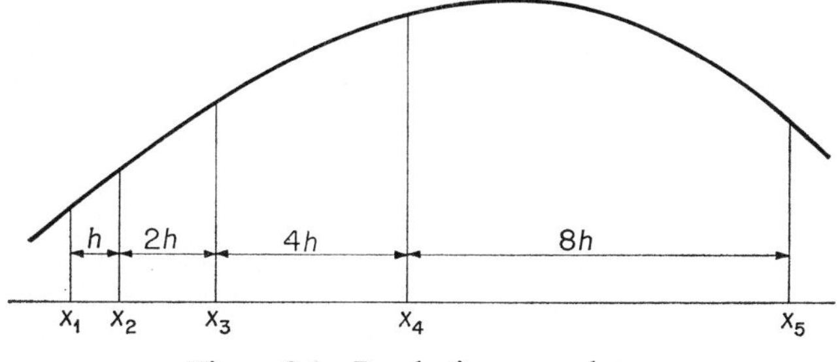

Figure 2.1 Bracketing procedure.

Key decisions must be made at the start of the calculation, the values of a, h, X and the direction of search. Often the search direction cannot be decided on other evidence and both directions must be tried. It is usually better to err on the small side for h and the large side for X, the doubling procedure soon lengthens the interval and soon reaches X. This method, of course, is not foolproof but is found to work well (see problems 1 and 2).

Example 18 Bracket the maximum of $f = \tanh x/(1 + x^2)$.

Now

$$f' = \frac{\operatorname{sech}^2 x}{1 + x^2} - \frac{2x \tanh x}{(1 + x^2)^2}.$$

Since $f' > 0$ when $x = 0$ and since $f \to 0$ as $x \to \infty$ it is safe to assume that the maximum occurs for $x > 0$. Choose $h = 0.1$, $a = 0.1$, $X = 100$.

Table 2.1 Bracketing procedure.

x	0.1	0.2	0.4	0.8	1.6
f	0.098 68	0.189 78	0.327 54	0.404 90	0.258 90
f'	0.960 72	0.851 09	0.511 73	−0.054 14	−0.190 44

Using function values only the bracket for the maximum is (0.4, 1.6), while if the derivative is also used the bracket is (0.4, 0.8).

2.2.2 POLYNOMIAL APPROXIMATION

Once a bracket has been obtained for the extremum it is then required to obtain the extremum to any required accuracy. One simple way of doing this is to use the information obtained by the bracketing procedure directly and approximate this information by a polynomial (see problem 3). Suppose $f_1 = f(z_1), f_2 = f(z_2), f_3 = f(z_3)$ are known and $f_2 > f_1, f_3$ so that (z_1, z_3) brackets the maximum of the function f. A quadratic approximation for f can be written

$$f = (z - z_2)(z - z_3)F_1 + (z - z_3)(z - z_1)F_2 + (z - z_1)(z - z_2)F_3$$

where $F_1 = f_1/(z_1 - z_2)(z_1 - z_3)$, etc. This quadratic has a maximum at $f'(z) = 0$ or at

$$z^* = \tfrac{1}{2}(z_1 + z_2 + z_3) - \frac{\tfrac{1}{2}(z_1F_1 + z_2F_2 + z_3F_3)}{F_1 + F_2 + F_3}. \tag{2.1}$$

Thus an estimate, $f^* = f(z^*)$, for the extremum is now available. A new calculation can be commenced for the maximum by identifying the 'best' bracket from the values z_1, z_2, z^*, z_3, relabelling appropriately and using (2.1) again.

There are many variants of this algorithm, these depend mainly on the expense of evaluating the function. If the function values are easy to obtain one alternative is to choose the maximum from f_2 and f^*. If $f_2 \geq f^*$ keep z_2 as the best value but if $f_2 < f^*$ put $z_2 = z^*$. A new calculation is then performed by putting $h = \tfrac{1}{4}(z_3 - z_1)$ and then using new values $z_1 = z_2 - h$, $z_3 = z_2 + h$. The formula (2.1) takes the particularly simple form

$$z^* = z_2 + \frac{\tfrac{1}{2}h(f_1 - f_2)}{f_1 - 2f_2 + f_3} \tag{2.2}$$

and provides a very suitable hand computation method.

Example 18(*a*) Use the quadratic approximation algorithm on the function

$$f = \frac{\tanh z}{1 + z^2}.$$

From the previous bracketing procedure in section 2.2.1 (0.4, 1.6) brackets the maximum.

z	0.4	1.0	1.6	$z(\text{new}) = 1 + \tfrac{1}{2}.0.6.\dfrac{0.068\,64}{-0.175\,16} = 0.88$
f	0.327 54	0.387 54	0.258 90	$f(0.88) = 0.398\,12.$

Now choose $z_2 = 0.88$, $h = 0.3$.

z	0.58	0.88	1.18	$z(\text{new}) = 0.77$
f	0.391 10	0.398 12	0.345 87	$f(0.77) = 0.406\ 15.$

The best value to date is $z = 0.77, f = 0.406\ 15$. Continuing the calculation on the computer gives $z = 0.741\ 60, f = 0.406\ 53$ to five figure accuracy with 34 function evaluations.

Working directly with the values $z = 0.4, 0.8, 1.6$, obtained from the bracketing in section 2.2.1, an implementation of (2.1) on the computer gives the five figure accuracy with 17 function evaluations.

If the *derivative* is also available, immediate advantage is obtained as can be observed from example 18 in section 2.2.1 where a much more accurate bracket was obtained. Suppose a maximum of f is bracketed by (x_1, x_2) and hence f_1, f'_1, f_2, f'_2 are known. Since there are now four pieces of information known it is possible to approximate the function in the interval by a cubic.

$$f(x) = ax^3 + bx^2 + cx + d \tag{2.3}$$

$$f'(x) = 3ax^2 + 2bx + c. \tag{2.4}$$

Putting the known information into (2.3) and (2.4) the result can be summarized in the matrix equation (2.5)

$$\begin{bmatrix} f_1 \\ f_2 \\ f'_1 \\ f'_2 \end{bmatrix} = \begin{bmatrix} x_1^3 & x_1^2 & x_1 & 1 \\ x_2^3 & x_2^2 & x_2 & 1 \\ 3x_1^2 & 2x_1 & 1 & 0 \\ 3x_2^2 & 2x_2 & 1 & 0 \end{bmatrix} \begin{bmatrix} a \\ b \\ c \\ d \end{bmatrix}, \tag{2.5}$$

which can be solved for a, b, c, d. The maximum of the approximating cubic is given by $f'(x) = 0$ from equation (2.4) or

$$x^* = \frac{-b \pm (b^2 - 3ac)^{\frac{1}{2}}}{3a} \tag{2.6}$$

and the sign is chosen to make $x_1 < x^* < x_2$. The gradients f'_1, f'_*, f'_2 are compared and the two new points are chosen to give gradients of opposite sign. The same procedure can then be repeated until sufficient accuracy is obtained.

This method works extremely well for most functions and is very well used. It is, however, not particularly suitable for a hand computation since a matrix equation (2.5) must be solved and a square root taken in (2.6) (see problem 5). On a computer these problems are straightforward and such an algorithm is usually available in most programme libraries.

Example 18(*b*) Use the cubic approximation algorithm to find the maximum of $f = \tanh x/(1 + x^2)$.

From section 2.2.1, example 18, the maximum is bracketed by (0.4, 0.8).

x	f	f'
0.4	0.327 54	0.511 73
0.8	0.404 90	−0.054 14

Solve (2.5); $a = 0.442\,4$, $b = -1.504$, $c = 1.502$
From (2.6); $x^* = 0.744$.

This new value gives $f_* = 0.406\,53$, $f'_* = -0.002\,44$, which is already better than the quadratic approximation, example 18(a). Since $f'(0.4)$ and $f'(0.744)$ have opposite signs these are chosen for the next approximation.

x	f	f'
0.4	0.327 54	0.511 73
0.744	0.406 53	−0.002 44

If the solution of (2.5) and (2.6) is repeated the new value gives $x^* = 0.741\,7$,
$f_* = 0.406\,53, f'_* = -0.000\,11$.

The calculation proceeds on the computer to give $f' = 0$ to 5 decimal places at the next iteration with $x = 0.741\,60, f = 0.406\,53$.

2.2.3 FIBONACCI TYPE SEARCH

This class of methods is often called the class of direct search methods and is concerned with optimizing when the derivative is *not* known. Once a bracket has been obtained the aim is to progressively reduce the length of the bracket until it is less than a prescribed limit.

Suppose (a_1, a_2) brackets a required *maximum* of the function $f(x)$. The points a_3, a_4 are symmetrically placed in this interval, so that

$$\left.\begin{aligned} a_3 &= (1 - \alpha)a_1 + \alpha a_2, \\ a_4 &= \alpha a_1 + (1 - \alpha)a_2, \end{aligned}\quad 0 < \alpha < \tfrac{1}{2}\right\} \qquad (2.7)$$

and this division is illustrated in figure 2.2. Calculate $f_i = f(a_i)$, $i = 1, 2, 3, 4$; then *either* $f(a_4) > f(a_3)$ and it is now assumed that (a_3, a_2) brackets the

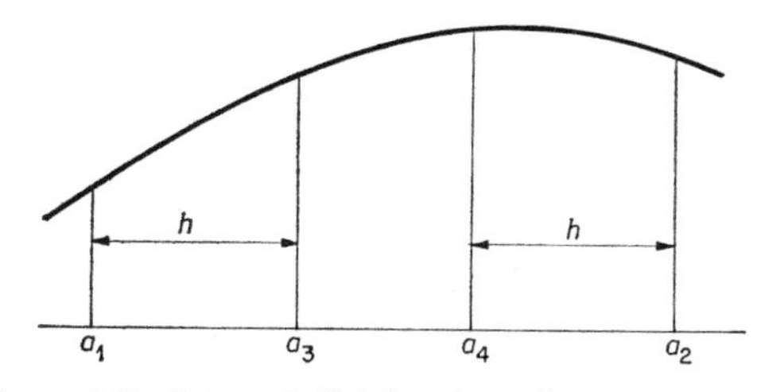

Figure 2.2 Interval division for Fibonacci search.

maximum *or* $f(a_3) \geq f(a_4)$ in which case (a_1, a_4) is assumed to be the bracket. Since the number of function evaluations must be reduced to a minimum, it would be very convenient to use the three remaining values in a further symmetrical division of the reduced interval.

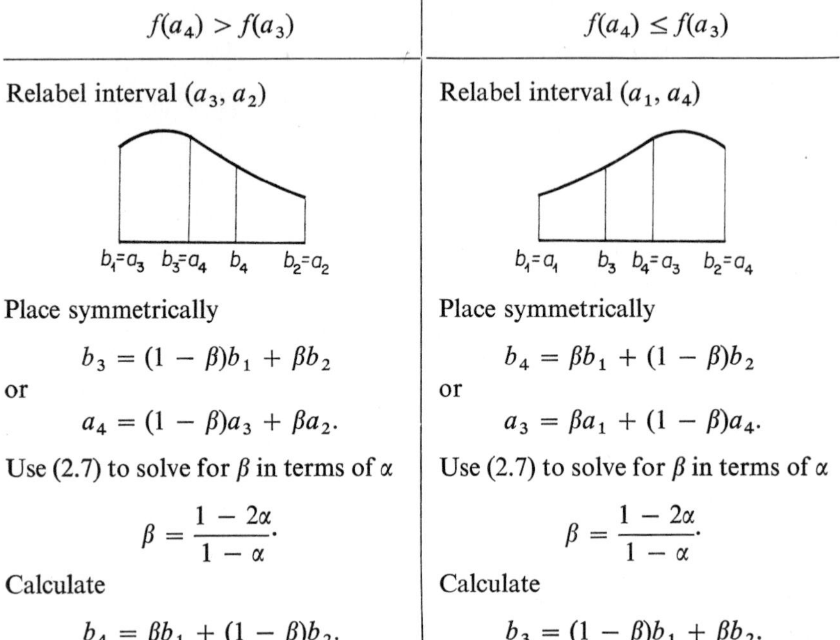

$f(a_4) > f(a_3)$	$f(a_4) \leq f(a_3)$
Relabel interval (a_3, a_2)	Relabel interval (a_1, a_4)
$b_1 = a_3$, $b_3 = a_4$, b_4, $b_2 = a_2$	$b_1 = a_1$, b_3, $b_4 = a_3$, $b_2 = a_4$
Place symmetrically	Place symmetrically
$b_3 = (1 - \beta)b_1 + \beta b_2$	$b_4 = \beta b_1 + (1 - \beta)b_2$
or	or
$a_4 = (1 - \beta)a_3 + \beta a_2.$	$a_3 = \beta a_1 + (1 - \beta)a_4.$
Use (2.7) to solve for β in terms of α	Use (2.7) to solve for β in terms of α
$\beta = \frac{1 - 2\alpha}{1 - \alpha}.$	$\beta = \frac{1 - 2\alpha}{1 - \alpha}.$
Calculate	Calculate
$b_4 = \beta b_1 + (1 - \beta)b_2.$	$b_3 = (1 - \beta)b_1 + \beta b_2.$

In both cases β has the same value which is determined uniquely by α. The method can be continued in precisely the same way, successive symmetric interval divisions being performed until the length of the interval is less than the required tolerance. The sequence of fractions $\alpha, \beta, \ldots$ can be labelled more conveniently as $\alpha_1, \alpha_2, \alpha_3, \ldots$ and they satisfy the recurrence relation

$$\alpha_{n+1} = \frac{1 - 2\alpha_n}{1 - \alpha_n} \tag{2.8}$$

The basic choice is how to satisfy (2.8) in the most convenient manner. Two methods will be described and compared. The first puts $\alpha = \alpha_1 = \alpha_2 = \alpha_3 = \cdots$ and solving (2.8) gives $\alpha = \frac{1}{2}[3 - (5)^{\frac{1}{2}}] = 0.382$, this value being used consistently in the interval divisions. This method is often called the method of *Golden section* (see problem 6). The second method of satisfying (2.8) uses the *Fibonacci* numbers and works with a *given* number, N, of interval divisions. Fibonacci, who was a 13th century monk, considered the sequence of numbers satisfying

$$F_0 = F_1 = 1, \qquad F_n = F_{n-1} + F_{n-2} \qquad n \geq 2;$$

the first few are 1, 1, 2, 3, 5, 8, 13, 21, 34, 55, ... Taking

$$\alpha_i = \frac{F_{N-i-1}}{F_{N-i+1}}$$

$$\frac{1 - 2\alpha_i}{1 - \alpha_i} = \frac{F_{N-i+1} - 2F_{N-i-1}}{F_{N-i+1} - F_{N-i-1}} = \frac{F_{N-i-2}}{F_{N-i}} = \alpha_{i+1}$$

and hence (2.8) is satisfied. The sequence of α's for this method is $\alpha_1 = F_{N-2}/F_N$, $\alpha_2 = F_{N-3}/F_{N-1}, \ldots, \alpha_{N-1} = F_0/F_2$. In the final division with $\alpha_{N-1} = F_0/F_2 = \frac{1}{2}$ it is seen that the two end points and the *mid*point have been selected. These three values have already been calculated and it is usual to evaluate the function $f(x)$ at a point close to the midpoint to decide which half to choose as the final interval.

By considering the division illustrated in figure 2.2 it may be seen that if the initial interval has length L then after the first reduction the length is $(1 - \alpha_1)L$ and after the second reduction $(1 - \alpha_1)(1 - \alpha_2)L$ etc. After $(N - 1)$ steps

$$\text{golden section length} = (1 - \alpha)^{N-1}L = \{\tfrac{1}{2}[-1 + (5)^{\frac{1}{2}}]\}^{N-1}L$$

$$\text{Fibonacci length} = \frac{F_{N-1}}{F_N}\frac{F_{N-2}}{F_{N-1}} \cdots \frac{F_0 L}{F_2} = \frac{L}{F_N}.$$

Since $F_N \to \{\frac{1}{2}[1 + (5)^{\frac{1}{2}}]\}^{N+1}/(5)^{\frac{1}{2}}$ as $N \to \infty$ (see problem 7)

$$\frac{\text{golden section length}}{\text{Fibonacci length}} = \frac{1}{(5)^{\frac{1}{2}}}\left(\frac{1 + (5)^{\frac{1}{2}}}{2}\right)^2 = 1.17.$$

Thus for large N the Fibonacci search gives a 17% better result than the golden section.

Example 18(*c*) Apply the golden section and Fibonacci searches to $f = \tanh x/(1 + x^2)$.

Golden section

$$a_3 = 0.618a_1 + 0.382a_2$$

$$a_4 = 0.382a_1 + 0.618a_2$$

The final interval after 14 function evaluations (including the initial four) is (0.738 1, 0.744 1), that is an accuracy of about 1% with a best value of

Table 2.2 Golden section search.

	a	f	a	f	a	f	a	f	a	f
1	0.4	0.327 5	0.4	0.327 5	0.4	0.327 5	0.575	0.390 1	0.683	0.404 7
3	0.858	0.400 4	*0.683*	0.404 7	*0.575*	0.390 1	0.683	0.404 7	0.750	0.406.5
4	1.142	0.353 7	0.858	0.400 4	0.683	0.404 7	*0.750*	0.406 5	*0.791*	0.405 3
2	1.6	0.258 9	1.142	0.353 7	0.858	0.400 4	0.858	0.400 4	0.858	0.400 4

$f = 0.406\ 53$ at $a = 0.740\ 4$. The function value is certainly accurate to five decimal places since f takes this same value over a whole range of values in the final bracket.

Fibonacci

$$a_3 = (1 - \alpha_i)a_1 + \alpha_i a_2, \qquad a_4 = \alpha_i a_1 + (1 - \alpha_i)a_2$$

$$\alpha_1 = \frac{89}{233}, \qquad \alpha_2 = \frac{55}{144}, \ldots.$$

The first few calculations give:

Table 2.3 Fibonacci search.

	a	f	a	f	a	f	a	f	a	f
1	0.4	0.327 5	0.4	0.327 5	0.4	0.327 5	0.575	0.390 1	0.683	0.404 7
3	0.858	0.400 4	*0.683*	0.404 7	*0.575*	0.390 1	0.683	0.404 7	0.750	0.406 5
4	1.142	0.353 8	0.858	0.400 4	0.683	0.404 7	*0.750*	0.406 5	*0.791*	0.405 3
2	1.6	0.258 9	1.142	0.353 8	0.858	0.400 4	0.858	0.400 4	0.858	0.400 4

Note that tables 2.2 and 2.3 are almost identical to the number of decimal places quoted. Eventually after a further four steps the values of a are $a_1 = 0.734\ 8$, $a_3 = a_4 = 0.739\ 9$, $a_2 = 0.745\ 1$. Since the two midpoints are identical the process terminates. A final calculation at $a = 0.740\ 0$ shows that after 14 function evaluations the final bracket is $(0.739\ 9, 0.745\ 1)$ with a best value of $f = 0.406\ 53$ at $a = 0.740\ 0$.

This method is a comparatively straightforward one to compute; it requires little storage and is reasonably efficient. In practical cases it is found to work very well but as with most of these algorithms it is not foolproof. For instance a large sharp peak between a_1 and a_3 of figure 2.2 would be missed. However, for a unimodal function (i.e. one with a single maximum or minimum) it can be proved that the method always works.

2.3 Several Variables, Derivatives not Available

2.3.1 Introduction

In problems where derivatives are not available it is usual that the function under consideration is complicated and little is known of its structure. The contours of the given function (see figure 2.3) are not known and not much can be deduced about them from the function itself. It is clear that if this is the case, the convergence of such methods will be very slow. This is certainly borne out by the study of single variable methods in the previous section.

The random search method described in section 2.2.1 is again applicable but is usually fairly inefficient. Another method that can be used is a *lattice search*. In a two variable problem illustrated in figure 2.3, a regular mesh is placed on the plane and the function evaluated at the mesh points. When the

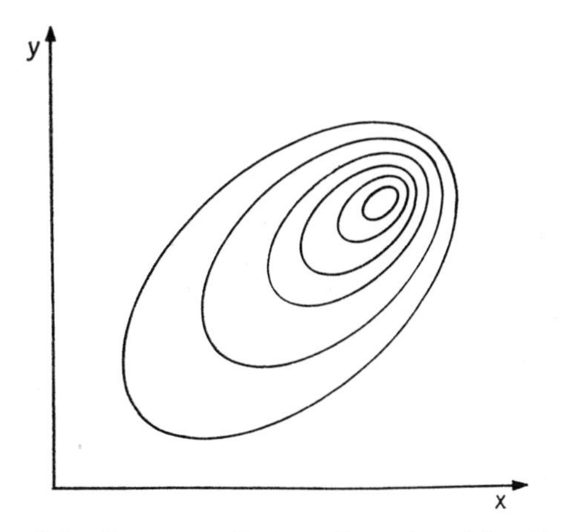

Figure 2.3 Contours of a two dimensional function.

'best' square has been obtained the mesh is then refined and the function evaluated at each of these new mesh points. The method then continues until a satisfactory tolerance is obtained. Since the function evaluations are the time consuming operation for this type of problem, a lattice search is usually wasteful. The mesh can, however, be used more efficiently, in particular the use of a triangular mesh leads to the *simplex method*. (This should not be confused with the simplex method used in linear programming in chapter 3.)

2.3.2 SIMPLEX METHOD

The basic method was first devised by Spendley and coauthors (1962) but it has since been modified extensively to make it a highly efficient routine. Only the original method will be described in detail since the variants of the method are discussed and compared by Parkinson and Hutchinson (1972). A computer package of one of these variants is usually available in a computer library.

The first step is to set up a regular simplex in an n-dimensional space, that is $(n + 1)$ points $\mathbf{x}_0, \mathbf{x}_1, \ldots, \mathbf{x}_n$ all equidistant from each other. In two dimensions a regular simplex is an equilateral triangle and in three dimensions a regular tetrahedron. If the minimum of a function $f(\mathbf{x})$ is required, at each vertex $f_i = f(\mathbf{x}_i)$ is evaluated and the vertices re-ordered so that $f_0 \geq f_1 \geq \cdots \geq f_n$. Since f_0 is the worst value the method tries to make a move as far away as possible from $\mathbf{x}_0$ by reflecting the vertex in the centroid of the opposite side of the simplex, i.e. to

$$\mathbf{x} = \mathbf{x}_0 + 2(\bar{\mathbf{x}} - \mathbf{x}_0),$$

where

$$\bar{\mathbf{x}} = \frac{1}{n}(\mathbf{x}_1 + \mathbf{x}_2 + \cdots + \mathbf{x}_n).$$

A new simplex $\mathbf{x}, \mathbf{x}_1, \ldots, \mathbf{x}_n$ has been formed and computing the value $f = f(\mathbf{x})$ two possible cases arise: $f < f_1$ or $f \geq f_1$. In the first case f is not the worst vertex and by reordering the vertices again the same reflection procedure can be applied to the new simplex. If, on the other hand, $f \geq f_1$ then $\mathbf{x}$ gives the worst value of the new simplex so reapplying the reflection method to the new simplex would just reproduce the old. To get out of this loop a new strategy is employed; the next largest vertex of the new simplex, $\mathbf{x}_1$, is chosen and the reflection procedure implemented on it. If the strategy is a success the method proceeds as before. If the strategy fails the vertex $\mathbf{x}_2$ is selected for reflection. The cycle then continues similarly; if a success is obtained return to the original method, if a failure try the next largest vertex.

Eventually as successive reflections approach the minimum of the function the stage is reached when the above methods fail to make progress since the size of the simplices is too large. Computer experiments have shown that if a vertex occurs in more than $M = 1.65n + 0.05n^2$ successive simplices then the size of the simplex is too large. To overcome this difficulty the next operation is to shrink the length of the sides of the simplex by a factor of $\frac{1}{2}$. The best available point $\mathbf{x}_n$ is kept and the other points computed as $\frac{1}{2}(\mathbf{x}_n + \mathbf{x}_i)$, $i = 0, 1, \ldots, n - 1$. Once the new points have been reordered the whole process can be recommenced.

The three basic steps are:

(1) Reflect largest vertex in the opposite side.

(2) If (1) fails reflect in the next largest vertex of the new simplex.

(3) If a vertex occurs in M successive simplices, halve the sides of the simplex.

To illustrate these points consider example 16.

Table 2.4 Simplex method.

label	1	2	3	4	5	6	7	8
point	(0, 0)	(0, 0.5)	(0.433, 0.25)	(0.433, 0.75)	(0.866, 0.5)	(0.866, 1)	(1.3, 0.75)	(1.3, 1.25)
$\overline{V}$	9.79	7.32	4.86	3.68(*S*)	2.39(*S*)	2.15(*S*)	2.54(*F*)	3.23(*F*)
action	–	–	(1)	(1)	(1)	(1)	(2)	(2)
new simplex	–	–	2, 3, 4	3, 4, 5	4, 5, 6	5, 6, 7	6, 7, 8	6, 8, 9

Example 16 This problem reduced to minimizing

$$\bar{V} = (\bar{O}\bar{P} - \tfrac{1}{2})^2 + 2(\bar{B}\bar{P} - \tfrac{1}{2})^2 + 3(\bar{C}\bar{P} - \tfrac{1}{2})^2$$

where

$$X = \frac{x}{a}, \qquad Y = \frac{y}{a}, \qquad \bar{V} = \frac{V}{\lambda a},$$

$$\bar{O}\bar{P}^2 = X^2 + Y^2, \qquad \bar{B}\bar{P}^2 = X^2 + (Y - 1)^2,$$
$$\bar{C}\bar{P}^2 = (X - 2)^2 + (Y - 1)^2.$$

This is a non-dimensional form of original problem. In table 2.4 the simplex method is followed, S denoting success and F denoting fail. The next point, labelled 9, also gives a fail. Point 6 has now occurred in four successive simplices so the sides are halved and the method proceeds (table 2.5).

Table 2.5 Simplex method after shrinking.

label	6	10	11	12	13
point	(0.866, 1)	(0.866, 1.25)	(1.08, 1.125)	(1.08, 0.875)	(0.866, 0.75)
$\bar{V}$	2.15	2.68	2.36	2.03	2.05
action	–	–	(1)	(1)	(1)
new simplex	–	–	6, 12, 11	6, 13, 12	etc

For the two dimensional example above the procedure is illustrated in figure 2.4.

The major improvement of this method over a lattice search is that f is evaluated at selected vertices only. Although the method is more efficient than the lattice search, recent variants greatly improve the basic method by

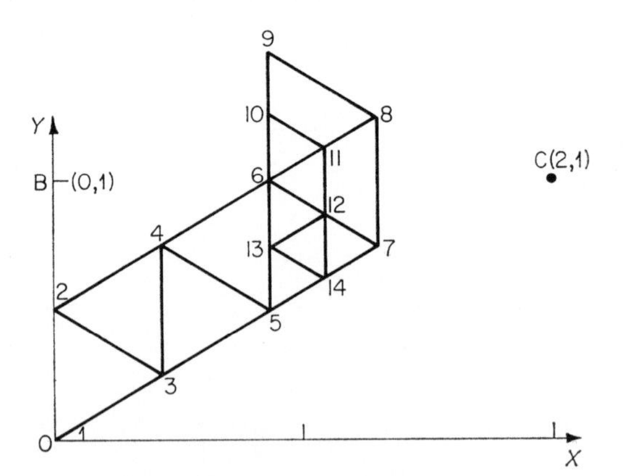

Figure 2.4 Simplex method, example 16.

removing the restriction of the regularity of the simplex; see Nelder and Mead (1965). The additional flexibility is obtained using the reflection formula

$$\mathbf{x} = \mathbf{x}_0 + \lambda(\bar{\mathbf{x}} - \mathbf{x}_0).$$

The value chosen by the basic method is $\lambda = 2$, while the modified method expands, $\lambda > 2$, if a success is recorded or contracts, $\lambda < 2$, for a failure. The detailed choice of λ and the implementation of the method can be found in the original article or in Parkinson and Hutchinson (1972).

2.3.3 ALTERNATING VARIABLE METHOD

This is perhaps the most straightforward of the several variable methods but suffers from this simplicity by being very slow. Each variable is chosen in turn, all the others are kept constant and the extremum is obtained by one of the single variable search methods, section 2.2.2 or 2.2.3. This method is best summarized by a flow diagram.

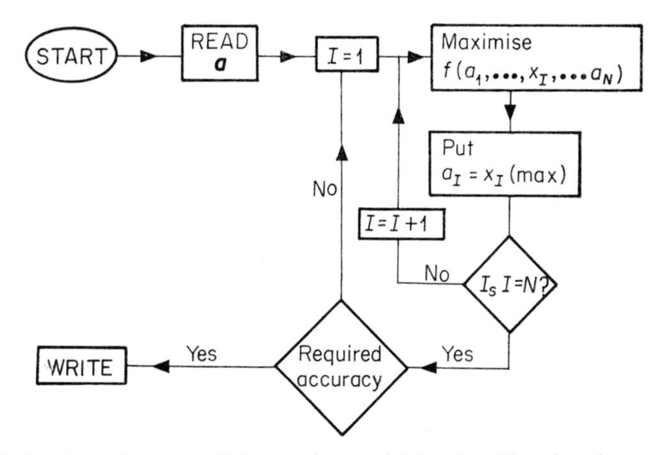

Figure 2.5 Flow diagram of alternating variable algorithm for the maximum of $f(x_1, x_2, \ldots, x_N)$.

2.3.4 ROSENBROCK'S MODIFICATION

Rosenbrock (1960) modified the alternating variable method in two ways to produce one of the most robust methods available for optimization when the derivatives are not available. The first of the modifications is to avoid the single variable optimization for each direction in turn. Instead a step of predetermined length is taken in each direction and these step lengths are modified after each calculation. If a step is taken in the '1-direction' and a better result is obtained, this is considered a 'good' direction and the step is lengthened (by a factor of 3 usually) for the next exploration of the 1-direction. If a 'worse' result is obtained then a shorter step (usually halved) in the opposite direction is chosen for the next search in the 1-direction. This procedure is followed as each of the variables are considered in turn.

The second modification is to recognize that the alternating variable method takes a large number of very small steps and then to try to avoid this by realigning the axes every so often. The axes are reoriented so that the first axis is along the *most successful overall* direction, the second axis along the next most successful direction and so on. If this can be achieved then considerable improvement can be expected. The real question, however, is to decide when and how to change the axes. The first question can only be answered by experience and it is found that the most useful criterion is to change axes when a 'good' *followed by a* 'bad' result has been obtained in *each* of the N directions, for an N variable problem. The question of how to change axes requires a very important orthogonalization process used extensively throughout analysis and algebra.

GRAM–SCHMIDT ORTHOGONALIZATION PROCESS

Given N linearly independent vectors $\mathbf{p}_1, \ldots, \mathbf{p}_N$ construct from them N orthonormal vectors $\mathbf{q}_1, \ldots, \mathbf{q}_N$. Thus the $\mathbf{q}_i$ must satisfy

$$\mathbf{q}_i \cdot \mathbf{q}_j = \delta_{ij} \qquad \text{all } i, j.$$

The method is as follows.

Choose

$$\mathbf{q}_1 = \frac{\mathbf{p}_1}{|\mathbf{p}_1|} \quad \text{and hence} \quad \mathbf{q}_1 \cdot \mathbf{q}_1 = 1.$$

Choose

$$\mathbf{q}_2 = a\mathbf{p}_1 + b\mathbf{p}_2 = \alpha\mathbf{q}_1 + \beta\mathbf{p}_2$$

and try to find α, β to satisfy

$$0 = \mathbf{q}_1 \cdot \mathbf{q}_2 = \alpha\mathbf{q}_1 \cdot \mathbf{q}_1 + \beta\mathbf{q}_1 \cdot \mathbf{p}_2 = \alpha + \beta(\mathbf{q}_1 \cdot \mathbf{p}_2),$$

hence

$$\mathbf{q}_2 = \beta[\mathbf{p}_2 - (\mathbf{q}_1 \cdot \mathbf{p}_2)\mathbf{q}_1] = \beta\mathbf{r}_2.$$

Now satisfy

$$\mathbf{q}_2 \cdot \mathbf{q}_2 = 1$$

so

$$\mathbf{q}_2 = \frac{\mathbf{r}_2}{|\mathbf{r}_2|}.$$

Choose

$$\mathbf{q}_3 = a\mathbf{p}_1 + b\mathbf{p}_2 + c\mathbf{p}_3 = \alpha\mathbf{q}_1 + \beta\mathbf{q}_2 + \gamma\mathbf{p}_3,$$

and try to find α, β, γ to satisfy

$$0 = \mathbf{q}_1 \cdot \mathbf{p}_3 = \alpha + \gamma(\mathbf{q}_1 \cdot \mathbf{p}_3)$$

$$0 = \mathbf{q}_2 \cdot \mathbf{p}_3 = \beta + \gamma(\mathbf{q}_2 \cdot \mathbf{p}_3)$$

and hence

$$\mathbf{q}_3 = \gamma[\mathbf{p}_3 - (\mathbf{q}_1 \cdot \mathbf{p}_3)\mathbf{q}_1 - (\mathbf{q}_2 \cdot \mathbf{p}_3)\mathbf{q}_2] = \gamma\mathbf{r}_3.$$

The normalization follows

$$\mathbf{q}_3 = \frac{\mathbf{r}_3}{|\mathbf{r}_3|}.$$

This process clearly continues and the $\mathbf{q}_1, \ldots, \mathbf{q}_N$ can be constructed to satisfy the required orthogonality conditions. The detailed proofs of independence, etc can be found in Mirsky (1955).

Example 19 Construct an orthonormal set from

$$(2, 0, 0, 0), \qquad (1, 1, 1, 0), \qquad (0, 1, 1, 1), \qquad (0, 0, 1, 2).$$

Now

$$\mathbf{q}_1 = (1, 0, 0, 0)$$

$$\mathbf{r}_2 = (1, 1, 1, 0) - 1(1, 0, 0, 0) = (0, 1, 1, 0)$$

$$\mathbf{q}_2 = \frac{1}{(2)^{\frac{1}{2}}}(0, 1, 1, 0)$$

$$\mathbf{r}_3 = (0, 1, 1, 1) - 0(1, 0, 0, 0) - 1(0, 1, 1, 0) = (0, 0, 0, 1)$$

$$\mathbf{q}_3 = (0, 0, 0, 1)$$

$$\mathbf{r}_4 = (0, 0, 1, 2) - 0\mathbf{q}_1 - \frac{1}{(2)^{\frac{1}{2}}}\mathbf{q}_2 - 2\mathbf{q}_3 = (0, -\tfrac{1}{2}, \tfrac{1}{2}, 0)$$

$$\mathbf{q}_4 = \frac{1}{(2)^{\frac{1}{2}}}(0, -1, 1, 0).$$

Return now to a more detailed look at Rosenbrock's method for finding the maximum of a function of three variables $f(x_1, x_2, x_3)$. Suppose at some stage of the calculation the current point is $\mathbf{a} = (a_1, a_2, a_3)$, the three search directions are $\mathbf{i}_1, \mathbf{i}_2, \mathbf{i}_3$ and the step lengths are s_1, s_2, s_3. Evaluate f at $\mathbf{a}, f_0 = f(a_1, a_2, a_3)$. Take a step in the $\mathbf{i}_k$ direction and evaluate $f_1 = f(\mathbf{a} + s_k\mathbf{i}_k)$.

$f_1 > f_0$	$f_1 \leq f_0$
Success	Failure
Replace **a** by $\mathbf{a} + s_k\mathbf{i}_k$ as best point	Keep **a** as best point
Replace f_0 by f_1 as best value	Keep f_0 as best value
Replace s_k by $3s_k$	Replace s_k by $-\frac{1}{2}s_k$

Advance k by $1(1 \to 2, 2 \to 3, 3 \to 1)$ and repeat the process. When a success and a fail has been logged in each of the three directions compute new orthogonal axes by the Gram–Schmidt process using $\mathbf{p}_1, \mathbf{p}_2, \mathbf{p}_3$ chosen as follows. Suppose the total progress made in the $\mathbf{i}_1, \mathbf{i}_2, \mathbf{i}_3$ directions are S_1, S_2, S_3 (all $\neq 0$) respectively then put

$$\mathbf{p}_1 = S_1\mathbf{i}_1 + S_2\mathbf{i}_2 + S_3\mathbf{i}_3$$

$$\mathbf{p}_2 = S_1\mathbf{i}_1 + S_2\mathbf{i}_2$$

$$\mathbf{p}_3 = S_1\mathbf{i}_1.$$

This choice is made so that $\mathbf{p}_1$, being the total progress made, will be retained. A typical example is illustrated for a two dimensional problem in figure 2.6.

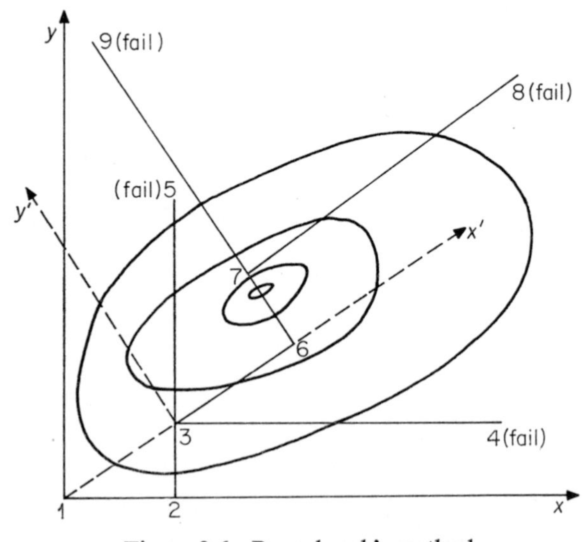

Figure 2.6 Rosenbrock's method.

Successive points are labelled 1, 2, 3, . . .; new axes are computed at 3. A numerical example to illustrate the method is as follows.

Table 2.6 Rosenbrock method.

θ_1	0	.1	0	0	−.05	0	0	.025	.025	.025	.1	.1	.1
θ_2	−.8	−.8	−.7	−.8	−.8	−.85	−.85	−.85	−1.0	−1.0	−1.0	−1.45	−1.45
θ_3	.8	.8	.8	.9	.9	.9	1.2	1.2	1.2	2.1	2.1	2.1	4.8
$2A$	.218	.215	.185	.255	.251	.275	.398	.402	.484	.844	.880	1.153	.016
S^* or F^*	–	F_1	F_2	S_3	F_1	S_2	S_3	S_1	S_2	S_3	S_1	S_2	F_3
s_1	.1	−.05	−.05	−.05	.025	.025	.025	.075	.075	.075	.225	.225	
s_2	.1	.1	−.05	−.05	−.05	−.15	−.15	−.15	−.45	−.45	−.45	−1.35	
s_3	.1	.1	.1	.3	.3	.3	.9	.9	.9	2.7	2.7	2.7	

* The suffix in this row indicates the current direction.

Example 20 Find the triangle with maximum area when the vertices of the triangle lie on the ellipse

$$x^2 + 4y^2 = 1.$$

Firstly the points on the ellipse can be taken as $\mathbf{r}_i = (\sin\theta_i, \frac{1}{2}\cos\theta_i)$, $i = 1, 2, 3$ and the area of the triangle can be calculated as

$$A = \tfrac{1}{2}|(\mathbf{r}_1 - \mathbf{r}_2) \times (\mathbf{r}_1 - \mathbf{r}_3)|.$$

The axes are now changed since each of the directions has shown a success and a fail. The current orthogonal directions in the $(\theta_1, \theta_2, \theta_3)$ space are (1, 0, 0), (0, 1, 0), (0, 0, 1) and the new directions are computed from

$$\mathbf{p}_1 = (0.1, -0.65, 1.3), \qquad \mathbf{p}_2 = (0.1, -0.65, 0), \qquad \mathbf{p}_3 = (0.1, 0, 0)$$

to give

$$\mathbf{q}_1 = (0.069, -0.449, 0.892), \qquad \mathbf{q}_2 = (0.135, -0.882, 0.451),$$
$$\mathbf{q}_3 = (0.989, 0.151, 0).$$

The new calculation starts:

Table 2.7 Rosenbrock method after change of axes.

θ_1	.1	.107	.121	.220	.241
θ_2	−1.45	−1.49	−1.58	−1.56	−1.69
θ_3	2.1	2.19	2.14	2.14	2.41
$2A$	1.153	1.193	1.218	1.225	1.289
S or F	–	S_1	S_2	S_3	S_1
s_1	.1	.3	.3	.3	.9
s_2	.1	.1	.3	.3	.3
s_3	.1	.1	.1	.3	.3

The method continues until a suitable convergence criterion is satisfied. To an accuracy of five significant figures in the θ values the method converged to $\theta_1 = 0.098\,923$, $\theta_2 = -1.996\,5$, $\theta_3 = 2.192\,3$ in about 200 steps. The problem is quite a testing one since the values $\theta_1 = \theta_2 + \frac{2}{3}\pi$, $\theta_3 = \theta_2 - \frac{2}{3}\pi$ give the same maximum value of $(3)^{3/2}/8$ for any value of θ_2. Even though the maximum is a weak maximum (as defined in section 1.2.1) a satisfactory solution was achieved. Starting at other values, similar convergence was achieved in about 200 steps and about 10 changes of axes.

On a more straightforward problem, the minimum of

$$f = x^2 + y^2 + 4(x - y + z - 1)^2$$

was found at (0, 0, 1) to an accuracy of five significant figures, from a starting value of $(-1, -1, -1)$ in 230 steps and 15 changes of axes.

2.4 Several Variables, Derivatives Available

2.4.1 Introduction

As described in the previous section very many hill climbing techniques for several variables reduce to a sequence of single variable searches. It is essential therefore that such searches are performed efficiently and hence the extensive discussion in section 2.2. One particular lesson that was learned from that section was that the derivative must be used if it is available. The same is equally true for many variable problems. In the quadratic example above for instance the Rosenbrock method required 230 function evaluations, while it could be expected that with the derivative available the minimum would be achieved in a very few steps. Indeed the gradient of a function provides a great deal of local information about the function under discussion.

The general approach outlined above uses the function value and the gradient at the current point of the calculation to estimate a 'best' direction of search. Having selected the direction, a single variable search for the minimum is then performed. This is usually done by the cubic search procedure of section 2.2.2. Once the minimum in the search direction is obtained a new search is initiated from the minimum point.

The basic problem in this approach is to decide what is meant by 'best'. One possible interpretation is to choose the direction of steepest descent (or ascent) at the current point of the calculation. To determine this direction consider a function $f(x_1, x_2, \ldots, x_n)$ at a point $\mathbf{x} = \mathbf{a}$. Take a small step $\mathbf{h}$ of fixed length L and select the direction of $\mathbf{h}$ to minimize the change in f. Thus it is required to minimize $f(a_1 + h_1, \ldots, a_n + h_n)$ subject to $h_1^2 + h_2^2 + \cdots + h_n^2 = L^2$. Using the Lagrange multiplier λ, the unconstrained optimum is required of

$$f^* = f + \lambda \sum h_i^2.$$

Hence

$$\frac{\partial f^*}{\partial h_k} = \frac{\partial f}{\partial h_k} + 2\lambda h_k = 0 \qquad (k = 1, \ldots, n),$$

and therefore

$$-2\lambda(h_1, \ldots, h_n) = \left(\frac{\partial f}{\partial h_1}, \ldots, \frac{\partial f}{\partial h_n}\right) = \nabla f$$

Thus the vector $\mathbf{h}$ is parallel to ∇f at $\mathbf{x} = \mathbf{a}$. It is easily shown that in the $(+\nabla f)$ direction the function increases and in the $(-\nabla f)$ direction the function decreases.

Locally the directions of steepest ascent and descent of the function f are ∇f and $-\nabla f$ respectively.

It should be recalled that ∇f is always perpendicular to the contours of f

(see problem 14) and this is illustrated in figure 2.7. It should also be noted that the local steepest direction is not necessarily the 'best' direction. Ideally this direction joins the current point to the top of the hill, but since it is usually impossible to obtain some compromise is necessary. One possible compromise is to use the steepest direction.

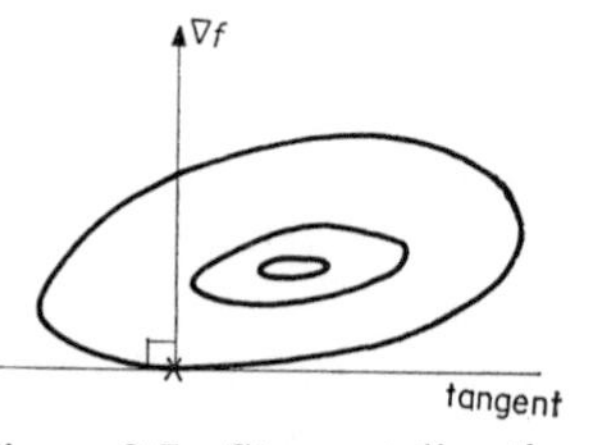

Figure 2.7 Steepest direction.

So far it has been assumed that only first derivatives are available. This may not be the case and higher derivatives provide an even more detailed picture of the contours of the function; better results can therefore be expected. To perform calculations using higher derivatives, it becomes convenient not to use the usual vector notation but to use matrix notation in the Taylor expansion of a function.

USE OF MATRIX NOTATION

Consider a function of two variables $f(x, y)$ at a point $P(a, b)$. At a neighbouring point $(a + h, b + k)$ Taylor's theorem states

$$f(a + h, b + k) = f(a, b) + hf_x + kf_y + \tfrac{1}{2}(h^2 f_{xx} + 2hkf_{xy} + k^2 f_{yy}) + \cdots,$$

where all functions $f_x, f_y, \ldots$ are evaluated at P. Now this can be written

$$f(a + h, b + k) = f_P + (h \;\; k)\begin{pmatrix} f_x \\ f_y \end{pmatrix} + \tfrac{1}{2}(h \;\; k)\begin{pmatrix} f_{xx} & f_{xy} \\ f_{xy} & f_{yy} \end{pmatrix}\begin{pmatrix} h \\ k \end{pmatrix} + \cdots$$

as can be verified by multiplying out the matrices. This method extends readily for many variables and for a general function $f(x_1, x_2, \ldots, x_n)$ the Taylor series of f about $P(a_1, a_2, \ldots, a_n)$ can be written in terms of

$$\mathbf{h} = \begin{bmatrix} h_1 \\ \cdot \\ \cdot \\ \cdot \\ h_n \end{bmatrix}, \quad \mathbf{D} = \begin{bmatrix} f_{x_1} \\ \cdot \\ \cdot \\ \cdot \\ f_{x_n} \end{bmatrix}, \quad \mathbf{J} = \begin{bmatrix} f_{x_1x_1} & f_{x_1x_2} & \cdots & f_{x_1x_n} \\ f_{x_2x_1} & f_{x_2x_2} & \cdots & f_{x_2x_n} \\ \cdot & & & \cdot \\ \cdot & & & \cdot \\ \cdot & & & \cdot \\ f_{x_nx_1} & f_{x_nx_2} & \cdots & f_{x_nx_n} \end{bmatrix}$$

as

$$f(a_1 + h_1, a_2 + h_2, \ldots, a_n + h_n) = f_P + \mathbf{h}^T\mathbf{D} + \tfrac{1}{2}\mathbf{h}^T\mathbf{J}\mathbf{h} + \cdots. \quad (2.9)$$

The matrix $\mathbf{J}$ is symmetric, $\mathbf{J}^T = \mathbf{J}$, and is usually called the *Jacobian* or *Hessian* matrix.

2.4.2 STEEPEST DESCENT AND NEWTON'S METHOD

The basis of these two straightforward methods has just been established in section 2.4.1. Suppose the minimum of $f(x_1, x_2, \ldots, x_n)$ is required then the steepest descent algorithm proceeds as follows:

(1) At the current point $\mathbf{a}_i$ compute $f(\mathbf{a}_i)$ and $\mathbf{h}_i = [\nabla f]_{\mathbf{x}=\mathbf{a}_i}$.

(2) Let $\mathbf{x} = \mathbf{a}_i - \lambda\mathbf{h}_i$. Find the minimum of the function of the single variable λ

$$F(\lambda) = f(\mathbf{a}_i - \lambda\mathbf{h}_i).$$

Since ∇f can be calculated at each point

$$\frac{dF}{d\lambda} = \sum_p \frac{\partial f}{\partial x_p}\frac{dx_p}{d\lambda} = -\nabla f \,.\, \mathbf{h}_i.$$

This can now be used to firstly bracket the minimum of F (section 2.2.1) and then obtain (section 2.2.2) the minimum of F at $\lambda = \lambda_{\min}$.

(3) Put

$$\mathbf{a}_{i+1} = \mathbf{a}_i - \lambda_{\min}\mathbf{h}_i \quad (2.10)$$

and return to step 1 unless the required convergence has been obtained.

The two main advantages of this method are that it is simple to apply and is very reliable. However, it has two major disadvantages. One is that step 2 is extremely tedious and slow even for a computer and there is a difficult decision of how accurately to perform each of the linear minimizations. The second disadvantage is similar to that described for the alternating variable method (section 2.3.3) that is, it takes a very large number of very small steps (see problem 15). In fact the two methods become almost identical since a steepest descent direction ($-\nabla f$) at P is always perpendicular to the contours through P. Since P is chosen by the minimization in step 2, the linear search direction is just the tangent to the contour at P. Thus all the successive directions are mutually perpendicular as in the alternating variable method.

Example 21(*a*) Minimize Rosenbrock's function

$$G = 10(x_1^2 - x_2)^2 + (1 - x_1)^2,$$

starting at (0, 0), by the steepest descent algorithm.

This function is an extremely awkward one since it has a long curved narrow valley with the minimum at (1, 1) in the valley.

$$G_{x_1} = 40x_1(x_1^2 - x_2) - 2(1 - x_1)$$

$$G_{x_2} = -20(x_1^2 - x_2).$$

Step 1 $\quad a_1 = (0, 0), \quad G = 1, \quad \mathbf{h}_1 = \nabla G = (-2, 0)$

Step 2 $\quad \mathbf{x} = (0, 0) - \lambda(-2, 0) = (2\lambda, 0)$

$$\frac{dG}{d\lambda} = 2G_{x_1} + 0G_{x_2}.$$

λ	0	0.1	0.3	
x_1	0	0.2	0.6	Approximate by a cubic, section 2.2.2
x_2	0	0	0	$\lambda_{\min} = 0.165$
G	1	0.656	1.456	$x_1 = 0.33, x_2 = 0$
$dG/d\lambda$	-4	-2.56	15.68	$G = 0.567\,5$, $dG/d\lambda = 0.195\,0$.

One further application of approximating by a cubic using the values for $\lambda = 0.1$ and $\lambda = 0.165$ gives $\lambda = 0.161$, $x_1 = 0.324$, $x_2 = 0$, $G = 0.567\,2$, $dG/d\lambda = 0.017\,0$.

Step 3 Taking this to be the best available λ the next iteration is started with these values.

Step 1 $\quad \mathbf{a}_2 = (0.324, 0), \quad G = 0.567\,2, \quad \mathbf{h}_2 = (0.009, -2.100)$
$\mathbf{x} = (0.324, 0) - \lambda(0.009, -2.100).$

Step 2

λ	0	0.1	
x_1	0.324	0.323	Approximation by a cubic gives
x_2	0	0.210	$\lambda = 0.049$, $x_1 = 0.323\,6$, $x_2 = 0.102\,9$
G	0.567 2	0.570 0	$G = 0.457\,5$, $dG/d\lambda = -0.063\,6$.
$dG/d\lambda$	-4.41	4.46	

The method continues with repeated applications of steps 1, 2 and 3.

Return now to *Newton's method* which assumes that the function may be approximated locally by its Taylor series up to the quadratic terms. Using matrix notation, equation (2.9) gives

$$f(a_1 + h_1, \ldots, a_n + h_n) = f_P + \mathbf{h}^T\mathbf{D} + \tfrac{1}{2}\mathbf{h}^T\mathbf{J}\mathbf{h} \tag{2.11}$$

where f_P, $\mathbf{D}$ and $\mathbf{J}$ are evaluated at the current point $P(a_1, \ldots, a_n)$ and are known.

The extremum of this approximating quadratic will provide $\mathbf{h}$ and hence a new estimate of the position of the extremum of the true function.

Let the column vector $\mathbf{E}_i$ be defined by $\mathbf{E}_i^T = (00 \ldots 100 \ldots 0)$ with the 1 in the ith place. Differentiating (2.11) with respect to h_i gives at the extremum

$$\frac{\partial f}{\partial h_i} = \mathbf{E}_i^T\mathbf{D} + \tfrac{1}{2}\mathbf{E}_i^T\mathbf{J}\mathbf{h} + \tfrac{1}{2}\mathbf{h}^T\mathbf{J}\mathbf{E}_i = 0.$$

But since

$$\mathbf{h}^T\mathbf{J}\mathbf{E}_i = (\mathbf{h}^T\mathbf{J}\mathbf{E}_i)^T = \mathbf{E}_i^T\mathbf{J}\mathbf{h}$$

this equation gives

$$0 = \mathbf{E}_i^T(\mathbf{D} + \mathbf{J}\mathbf{h})$$

and since this is true for each $\mathbf{E}_i$

$$0 = \mathbf{D} + \mathbf{J}\mathbf{h}$$

or the required column vector $\mathbf{h}$ can be calculated as

$$\mathbf{h} = -\mathbf{J}^{-1}\mathbf{D}. \tag{2.12}$$

Thus the extremum of the function is at

$$\begin{bmatrix} x_1 \\ \cdot \\ \cdot \\ \cdot \\ x_n \end{bmatrix} = \begin{bmatrix} a_1 \\ \cdot \\ \cdot \\ \cdot \\ a_n \end{bmatrix} - \mathbf{J}^{-1}\mathbf{D} \tag{2.13}$$

or for simplicity $\mathbf{x} = \mathbf{a} - \mathbf{J}^{-1}\mathbf{D}$.

The *Newton algorithm* now takes the form

(1) At the current point $\mathbf{a}_i$ evaluate f_i, $\mathbf{D}_i$, $\mathbf{J}_i$.
(2) Compute $\mathbf{a}_{i+1} = \mathbf{a}_i - \mathbf{J}_i^{-1}\mathbf{D}_i$ and return to (1) unless sufficient accuracy has been obtained.

This algorithm has two very distinct advantages, first it *avoids* the single variable search (i.e. no need to search in $-\mathbf{J}^{-1}\mathbf{D}$ direction) and secondly when it works the convergence is very rapid indeed since the method is second order with error of $0(\mathbf{h}^2)$. If the function is an exact quadratic the extremum is achieved in a single iteration, but in general the method will only work well if (2.11) is a good approximation to f. This implies geometrically that the local contours of the approximating quadratic reflect the overall shape of the maximum or minimum of f. This in turn implies algebraically that $\mathbf{J}$ is positive or negative definite in the respective cases. ($\mathbf{y}^T\mathbf{J}\mathbf{y} > 0$ for all $\mathbf{y}$ if positive definite.)

It can be deduced from the above paragraph that this method (although ideal for points sufficiently close to the extremum) is often extremely unreliable. This unreliability is usually produced by the approximating quadratic having a saddle point far distant from the required extremum. A second major disadvantage is that second derivatives, and hence **J**, are often difficult to compute and the matrix inversion, $\mathbf{J}^{-1}$, is time consuming. For these reasons Newton's method is not well used in this work and the adaptation of Davidon in section 2.4.3 has superseded it.

Example 21(*b*) Find the minimum of Rosenbrock's function

$$G = 10(x^2 - y)^2 + (1 - x)^2,$$

starting at $(-1, 1)$, using Newton's algorithm.

$$G_x = 40x(x^2 - y) - 2(1 - x) \qquad G_{xx} = 40(3x^2 - y) + 2$$

$$G_y = -20(x^2 - y) \qquad G_{xy} = -40x$$

$$G_{yy} = 20.$$

Step 1 $\quad \mathbf{a}_0 = \begin{pmatrix} -1 \\ 1 \end{pmatrix}, \quad G = 4, \quad \mathbf{D}_0 = \begin{pmatrix} -4 \\ 0 \end{pmatrix}, \quad \mathbf{J}_0 = \begin{pmatrix} 82 & 40 \\ 40 & 20 \end{pmatrix}.$

Step 2 $\quad \mathbf{J}_0^{-1} = \dfrac{1}{40}\begin{pmatrix} 20 & -40 \\ -40 & 82 \end{pmatrix}$

$$\mathbf{a}_1 = \begin{pmatrix} -1 \\ 1 \end{pmatrix} - \begin{pmatrix} 0.5 & 1 \\ -1 & 2.05 \end{pmatrix}\begin{pmatrix} -4 \\ 0 \end{pmatrix} = \begin{pmatrix} 1 \\ -3 \end{pmatrix}$$

Step 1 $\quad \mathbf{a}_1 = \begin{pmatrix} 1 \\ -3 \end{pmatrix}, \quad G = 160, \quad \mathbf{D}_1 = \begin{pmatrix} 160 \\ -80 \end{pmatrix}, \quad \mathbf{J}_1 = \begin{pmatrix} 242 & -40 \\ -40 & 20 \end{pmatrix}$

Step 2 $\quad \mathbf{J}_1^{-1} = \dfrac{1}{3240}\begin{pmatrix} 20 & 40 \\ 40 & 242 \end{pmatrix}$

and

$$\mathbf{a}_2 = \begin{pmatrix} 1 \\ 1 \end{pmatrix}, \quad G = 0.$$

Thus it so happens in this case that the minimum is achieved in two iterations despite a considerably worse value at the end of the first iteration.

2.4.3 DAVIDON'S METHOD (FIRST DERIVATIVES REQUIRED)

Following an original idea by Davidon (1959), see Fletcher and Powell (1963), a whole class of methods has been developed. They have proved to be the most successful of gradient methods and are now extensively used.

The detailed proofs of convergence of these methods need heavy matrix manipulation and will not be reproduced. However, the main ideas can be easily followed. Indeed the main idea is simple; use the best features of the Newton and steepest descent algorithms and avoid the worst. The implementation of this programme is less straightforward but as a start the best and worst features can be listed: Newton's method converges rapidly but is unreliable, requires second derivatives and a matrix inversion; the steepest descent algorithm is slow but always decreases the function value. Translating these comments, it appears that a method is required that starts like a steepest descent and finishes like a Newton method but only uses first derivatives. Thus the search direction should be slowly deflected from the steepest descent direction to the Newton direction as the iterations proceed.

Using matrix notation these directions can be written (2.10), (2.12)

$$\mathbf{a}_{i+1} = \mathbf{a}_i - \lambda_{\min}\mathbf{I}\mathbf{D}_i \qquad \text{Steepest descent}$$

$$\mathbf{a}_{i+1} = \mathbf{a}_i - \mathbf{J}^{-1}\mathbf{D}_i \qquad \text{Newton.}$$

For the minimum of a function a compromise between $(-\mathbf{I}\mathbf{D}_i)$ and $(-\mathbf{J}^{-1}\mathbf{D}_i)$ is chosen as $(-\mathbf{H}_i\mathbf{D}_i)$ so

$$\mathbf{a}_{i+1} = \mathbf{a}_i - \lambda_{\min}\mathbf{H}_i\mathbf{D}_i \qquad \text{Davidon.} \tag{2.14}$$

A sequence of matrices $\mathbf{H}_i$ must now be constructed to be *symmetric* (since $\mathbf{J}^{-1}$ is symmetric), *positive definite* (since $\mathbf{J}$ is positive definite at the extremum) and so that $\mathbf{H}_i \to \mathbf{J}^{-1}$ as $i \to \infty$. This programme is a difficult one but the key to the solution is taken from obtaining an estimate of $\mathbf{H}_{i+1}$ from a quadratic function

$$f = c + \mathbf{h}^T\mathbf{D}_i + \tfrac{1}{2}\mathbf{h}^T\mathbf{J}_i\mathbf{h}, \tag{2.15}$$

where $\mathbf{D}_i$ and $\mathbf{J}_i$ are the gradient vector and Jacobian respectively at the point $\mathbf{a}_i$. The gradient of this quadratic at a neighbouring point $\mathbf{p}_i$ can be computed from,

$$\mathbf{D}_{i+1} = \left(\frac{\partial f}{\partial h_1} \;\; \frac{\partial f}{\partial h_2} \;\; \cdots \;\; \frac{\partial f}{\partial h_n}\right)^T_{\text{at } \mathbf{p}_i}$$

$$= \mathbf{D}_i + \mathbf{J}_i\mathbf{p}_i$$

or

$$\mathbf{J}_i\mathbf{p}_i = (\mathbf{D}_{i+1} - \mathbf{D}_i) = \mathbf{y}_i, \tag{2.16}$$

where the $\mathbf{y}_i$ is defined as the difference of the gradients. Now if the matrix $\mathbf{H}$ is to behave eventually like $\mathbf{J}^{-1}$ then $\mathbf{HJ} = \mathbf{I}$. Pre-multiplying (2.16) by $\mathbf{H}_{i+1}$, it is required that

$$(\mathbf{H}_{i+1}\mathbf{J}_i)\mathbf{p}_i = \mathbf{p}_i = \mathbf{H}_{i+1}\mathbf{y}_i.$$

This is the best that can be asked of the **H**'s since assuming that an approximation $\mathbf{H}_i$ is known, it is required to update $\mathbf{H}_i$ to $\mathbf{H}_{i+1}$ while still trying to maintain $\mathbf{HJ} = \mathbf{I}$.

For the true function f under consideration, $\mathbf{a}_i$ is known, $\mathbf{D}_i$ can be computed and $\mathbf{H}_i$ is known. The search direction is chosen as $\mathbf{h} = -\mathbf{H}_i\mathbf{D}_i$ and the minimum is sought in this direction, say at B, $\mathbf{a}_{i+1}$, see figure 2.8. Writing $\mathbf{AB} = -\lambda_{\min}\mathbf{h} = \mathbf{p}_i$, the $\lambda_{\min}$ is obtained from a single variable minimization of $F(\lambda) = f(\mathbf{a}_i - \lambda\mathbf{h})$.

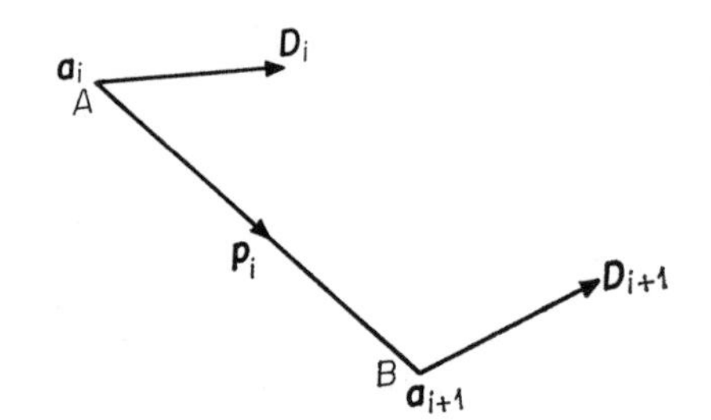

Figure 2.8 Successive points in Davidon's method.

Now that $\mathbf{a}_{i+1}$ is known the new $\mathbf{D}_{i+1}$ can be computed and hence $\mathbf{p}_i$ and $\mathbf{y}_i$. The only quantity that is required is an estimate of $\mathbf{H}_{i+1}$ in terms of the known vectors. Following the above deduction for a quadratic function, the basic equation to be satisfied is

$$\mathbf{H}_{i+1}\mathbf{y}_i = \mathbf{p}_i \tag{2.17}$$

for given $\mathbf{y}_i$ and $\mathbf{p}_i$. Writing

$$\mathbf{H}_{i+1} = \mathbf{H}_i + \mathbf{E}_i$$

then $\mathbf{E}_i$ can be treated as error and must satisfy (dropping the suffix i in the following)

$$\mathbf{Ey} = \mathbf{p} - \mathbf{Hy}. \tag{2.18}$$

Now a useful class of solutions to (2.18) was constructed by Huang (1970) and is a class which includes most of the previously known methods as special cases:

$$\mathbf{E} = -\frac{\mathbf{Hyu}^T}{\mathbf{u}^T\mathbf{y}} + \rho\frac{\mathbf{pv}^T}{\mathbf{v}^T\mathbf{y}}. \tag{2.19}$$

Substituting into (2.18) it may be seen that the equation is satisfied identically if the parameter $\rho = 1$. For the vectors **u** and **v** Huang chose

$$\mathbf{u} = \mathbf{p} + \alpha\mathbf{Hy}$$

$$\mathbf{v} = \mathbf{p} + \beta\mathbf{Hy}$$

thus providing a three parameter family (α, β, ρ) of formulae. He showed that, for all these formulae, the algorithm based on this updating of $\mathbf{H}$ gave convergence in n steps to the minimum of a quadratic function of n variables, provided that each linear search is done exactly.

By the construction of $\mathbf{E}$ it can be seen that all the Huang family are not necessarily symmetric or give positive definite updates for $\mathbf{H}$. This can be a serious disadvantage for many of the formulae and can lead to numerical difficulties. Various articles in Lootsma (1972) document these carefully. The best known of the formulae are (again dropping the suffix i in the right-hand sides)

$\rho = 1, \quad \alpha = \infty, \quad \beta = 0$ Davidon

$$\mathbf{H}_{i+1} = \mathbf{H} - \frac{\mathbf{H}\mathbf{y}\mathbf{y}^T\mathbf{H}}{\mathbf{y}^T\mathbf{H}\mathbf{y}} + \frac{\mathbf{p}\mathbf{p}^T}{\mathbf{p}^T\mathbf{y}} \tag{2.20}$$

$\rho = 1, \quad \alpha = \beta = -1$

$$\mathbf{H}_{i+1} = \mathbf{H} + \frac{(\mathbf{p} - \mathbf{H}\mathbf{y})(\mathbf{p} - \mathbf{H}\mathbf{y})^T}{(\mathbf{p} - \mathbf{H}\mathbf{y})^T\mathbf{y}} \tag{2.21}$$

$\rho = 1, \quad \alpha = 0, \quad \beta = -1\Big/\left(1 + \dfrac{\mathbf{y}^T\mathbf{H}\mathbf{y}}{\mathbf{p}^T\mathbf{y}}\right)$ Broyden-Fletcher

$$\mathbf{H}_{i+1} = \mathbf{H} + \frac{1}{\mathbf{p}^T\mathbf{y}}\left[\left(1 + \frac{\mathbf{y}^T\mathbf{H}\mathbf{y}}{\mathbf{p}^T\mathbf{y}}\right)\mathbf{p}\mathbf{p}^T - \mathbf{p}\mathbf{y}^T\mathbf{H} - \mathbf{H}\mathbf{y}\mathbf{p}^T\right]. \tag{2.22}$$

By construction these three formulae are symmetric and the first and third can be shown to provide positive definite $\mathbf{H}$'s (if the linear minimizations are performed exactly).

A major step in the calculations using a method of the type just described is the linear minimization. It is not clear how accurately these minimizations should be performed since it seems pointless to ask for minute accuracy early in a calculation. It can also be a time-consuming operation, very expensive in function evaluations. A great deal of effort therefore has been channelled into avoiding these linear searches. One approach has been to perform a 'weak search', just bracketing the minimum and then taking the best available value. It has been found, however, that this does not significantly reduce the number of function evaluations. A more useful idea that has been pursued is to choose the λ in (2.14) with only the proviso that the function value decreases at each step. This is not quite enough, however; additional conditions must be met to ensure that a sufficiently large decrease is obtained. The conditions can normally be met by successively trying $\lambda = 1, 0.1, 0.01, \ldots$ or by a single cubic interpolation. Eventually if $\mathbf{H} \to \mathbf{J}^{-1}$

then $\lambda = 1$ will be chosen so that, for most steps, only one function value will be required. Fletcher (1970) suggested this procedure using (2.22) for his updating formula and gives all the detailed conditions. Himmelblau (1972) has tested 15 algorithms on 15 test functions and concludes that the Fletcher method is the best in both speed and robustness followed by the Davidon formula (2.20).

Returning now to the construction of a Davidon type *algorithm*; its detailed implementation for the minimum of a function $f(x_1, \ldots, x_n)$, with derivatives that can be calculated, is as follows.

$\mathbf{a}_0$ and $\mathbf{H}_0$ (usually the unit matrix) are supplied and $\mathbf{D}_0$ and $f(\mathbf{a}_0)$ are then computed.

(*) At the ith stage f_i, $\mathbf{a}_i$, $\mathbf{D}_i$, $\mathbf{H}_i$ are known.

(1) Compute $\mathbf{a}_{i+1} = \mathbf{a}_i - \lambda \mathbf{H}_i \mathbf{D}_i$, where λ is chosen to minimize $F(\lambda) = f(\mathbf{a}_i - \lambda \mathbf{H}_i \mathbf{D}_i)$.
(2) Compute $\mathbf{D}_{i+1}$ at $\mathbf{a}_{i+1}$ and then $\mathbf{p}_i = \mathbf{a}_{i+1} - \mathbf{a}_i$, $\mathbf{y}_i = \mathbf{D}_{i+1} - \mathbf{D}_i$.
(3) Update $\mathbf{H}_i$ to $\mathbf{H}_{i+1}$ using (2.20), (2.21) or (2.22).
(4) Return to (*) unless convergence is complete.

As mentioned above the minimization of $F(\lambda)$ in step 1 can be avoided if (2.21) or (2.22) is used with a suitable choice of λ. To demonstrate these algorithms consider Box's function.

Example 22 Minimize the function

$$f = \sum_{\alpha} (e^{\alpha x} - e^{\alpha y} - e^{-\alpha} + e^{-10\alpha})^2,$$

where α takes the values 0.2, 0.4, 0.6, 0.8, 1.0.

Formula (2.20)
Step 1 Taking

$$\mathbf{a}_0 = \begin{pmatrix} 0 \\ 0 \end{pmatrix}, \quad f = 1.58, \quad \mathbf{D}_0 = \begin{pmatrix} -2.985 \\ 2.985 \end{pmatrix}, \quad \mathbf{H}_0 = \begin{pmatrix} 1 & 0 \\ 0 & 1 \end{pmatrix}$$

$$\mathbf{a}_1 = \begin{pmatrix} 0 \\ 0 \end{pmatrix} - \lambda \begin{pmatrix} 1 & 0 \\ 0 & 1 \end{pmatrix} \begin{pmatrix} -2.985 \\ 2.985 \end{pmatrix} = \begin{pmatrix} 2.985\lambda \\ -2.985\lambda \end{pmatrix}.$$

λ	0	0.1	0.2
x	0	0.298 5	0.598 0
y	0	−0.298 5	−0.598 0
f	1.58	0.580 7	1.368
$\dfrac{df}{d\lambda}$	−17.8	−1.396	18.10.

Cubic approximation using $\lambda = 0.1$ and 0.2 gives $\lambda = 0.159$ as the next approximation for the minimum. However, this value gives $f = 0.810\,2$ which is no improvement. Taking $\lambda = 0.1$ and 0.159, a further cubic approximation gives $\lambda = 0.106$ and $f = 0.575\,4$.

A more accurate value can be computed, but for the present purposes take the $\lambda = 0.106$ value as the best available.

Step 2

$$\mathbf{a}_1 = \begin{pmatrix} 0.316\,4 \\ -0.316\,4 \end{pmatrix}, \qquad f = 0.575\,4, \qquad \mathbf{D}_1 = \begin{pmatrix} 0.048\,17 \\ 0.173\,36 \end{pmatrix}$$

$$\mathbf{p} = \mathbf{a}_1 - \mathbf{a}_0 = \begin{pmatrix} 0.316\,4 \\ -0.316\,4 \end{pmatrix}, \qquad \mathbf{y} = \mathbf{D}_1 - \mathbf{D}_0 = \begin{pmatrix} 3.033 \\ -2.812 \end{pmatrix}.$$

Step 3

$$\mathbf{p}^T\mathbf{y} = 1.849, \qquad \mathbf{y}^T\mathbf{H}\mathbf{y} = 17.106,$$

$$0.540\,7\mathbf{p}\mathbf{p}^T = \begin{pmatrix} 0.054\,13 & -0.054\,13 \\ -0.054\,13 & 0.054\,13 \end{pmatrix},$$

$$0.058\,46(\mathbf{H}\mathbf{y})(\mathbf{H}\mathbf{y})^T = \begin{pmatrix} 0.537\,8 & -0.498\,6 \\ -0.498\,6 & 0.462\,3 \end{pmatrix},$$

$$\mathbf{H}_1 = \begin{pmatrix} 1 & 0 \\ 0 & 1 \end{pmatrix} - \begin{pmatrix} 0.537\,8 & -0.498\,6 \\ -0.498\,6 & 0.462\,3 \end{pmatrix} + \begin{pmatrix} 0.054\,1 & -0.054\,1 \\ -0.054\,1 & 0.054\,1 \end{pmatrix}$$

$$= \begin{pmatrix} 0.516\,3 & 0.444\,5 \\ 0.444\,5 & 0.591\,8 \end{pmatrix}.$$

Step 4 Convergence is clearly not complete so return to step 1.

The iteration proceeds in this way and eventually converges to the exact minimum at $(-1, -10)$. A programme of this method, which is available in most computing laboratories, gave convergence from $(0, 0)$ to $(-1, -10)$ in 15 iterations, taking successive values of x and y differing by less than 10^{-5} as the convergence criterion. With the same programme commencing the iterations at the points $(-10, -10)$ and $(0, -20)$ gave convergence in 24 and 25 iterations respectively. Starting at the point $(10, -10)$, however, the programme converged to point $(8.89, 8.89)$. At this point the gradients are small so presumably the method has found another local minimum or saddle point. The values of x and y differed by less than 10^{-5} at successive iterations and premature convergence ensued. This shows up two points; firstly the danger of relying on a single convergence criterion as in the programme used and secondly the advisability of starting a calculation from several different points.

Formula (2.22) without linear search for minimum.

Step 1

$$\mathbf{a}_0 = \begin{pmatrix} 0 \\ 0 \end{pmatrix}, \qquad f = 1.58, \qquad \mathbf{D}_0 = \begin{pmatrix} -2.985 \\ 2.985 \end{pmatrix}$$

$$\mathbf{H}_0 = \begin{pmatrix} 1 & 0 \\ 0 & 1 \end{pmatrix}, \qquad \mathbf{a}_1 = \begin{pmatrix} 2.985\lambda \\ -2.985\lambda \end{pmatrix}.$$

Taking $\lambda = 1$, gives a value $f = 400$ which does not decrease the function, so reduce to $\lambda = 0.1$. An acceptable reduction is obtained

$$\mathbf{a}_1 = \begin{pmatrix} 0.298\,5 \\ -0.298\,5 \end{pmatrix}, \qquad f = 0.580\,7, \qquad \mathbf{D}_1 = \begin{pmatrix} -0.169\,6 \\ 0.297\,2 \end{pmatrix}.$$

Step 2

$$\mathbf{p} = \mathbf{a}_1 - \mathbf{a}_0 = \begin{pmatrix} 0.298\,5 \\ -0.298\,5 \end{pmatrix}, \qquad \mathbf{y} = \mathbf{D}_1 - \mathbf{D}_0 = \begin{pmatrix} 2.815 \\ -2.688 \end{pmatrix}.$$

Step 3

$$\mathbf{p}^T\mathbf{y} = 1.643, \qquad \mathbf{y}^T\mathbf{H}\mathbf{y} = 15.15,$$

$$\mathbf{p}\mathbf{p}^T = \begin{pmatrix} 0.089\,10 & -0.089\,10 \\ -0.089\,10 & 0.089\,10 \end{pmatrix}, \qquad \mathbf{H}\mathbf{y}\mathbf{p}^T = \begin{pmatrix} 0.840\,3 & -0.840\,3 \\ -0.802\,4 & 0.802\,4 \end{pmatrix},$$

$$\mathbf{H}_1 = \begin{pmatrix} 1 & 0 \\ 0 & 1 \end{pmatrix} + 0.608\,8 \left[10.22 \begin{pmatrix} 0.089\,1 & -0.089\,1 \\ -0.089\,1 & 0.089\,1 \end{pmatrix} - \begin{pmatrix} 1.681 & -1.643 \\ -1.643 & 1.605 \end{pmatrix} \right] = \begin{pmatrix} 0.531\,0 & 0.447\,1 \\ 0.447\,1 & 0.577\,3 \end{pmatrix}.$$

Step 4 Convergence is not complete so return to step 1.

Again computing laboratories have this programme available and in this case very rapid convergence is obtained.

PROBLEMS

1. Bracket the maximum of the function

$$F = \frac{1}{(1-x)^2}\left(\log_e x + \frac{2(1-x)}{(1+x)}\right) \qquad \text{for } x \geq 1.$$

(*Note* This function is the lift force in a simple lubrication configuration.)

2. Use the bracketing procedure of section 2.2.1 for the maximum of the function

$$f(x) = \frac{1 + x}{1 + 2x^2}$$

starting at $x = 0$ with (i) initial step length 1, (ii) initial step length 0.1.
3. Show that an estimate for the extremum of a function $f(x)$ obtained by a quadratic approximation through $f_1 = f(a - h)$, $f_2 = f(a)$, $f_3 = f(a + 2h)$ is at the x-value

$$a + \tfrac{1}{2}h\left(\frac{4f_1 - 3f_2 - f_3}{2f_1 - 3f_2 + f_3}\right).$$

Apply this formula to the appropriate values given in the table of example 18 and compare with the results quoted in the text.
4. Use the quadratic approximation method in section 2.2.2 on the function F in problem 1.
5. Show that a, b, c in the matrix equation (2.5) can be found from the following.

Putting $D = 1/(x_2 - x_1)$, $F = D(f_2 - f_1)$, $G = D(f'_2 - f'_1)$, $H = D(F - f'_1)$, $K = D(G - 2H)$ then $a = K$, $b = H - a(x_2 + 2x_1)$, and finally $c = F - a(x_1{}^2 + x_1x_2 + x_2{}^2) - b(x_1 + x_2)$.

Use this and equation (2.6) to find a first estimate to the maximum of the function f in problem 2.
6. Let an interval AB be divided symmetrically at C, D. If $AC/AD = AD/AB$ show that C divides AB in the golden section ratio. (If this is done to each side of a rectangular picture, the lines joining opposite corresponding points are supposed to meet at the key points of the picture.)
7. Put $F_n = \alpha^n$ into the Fibonacci recurrence relation

$$F_0 = F_1 = 1, \qquad F_n = F_{n-1} + F_{n-2}, \qquad n \geq 2.$$

Deduce the possible values of α and hence show that

$$F_n \to \frac{\{\frac{1}{2}[1 + (5)^{\frac{1}{2}}]\}^{n+1}}{(5)^{\frac{1}{2}}} \qquad \text{as } n \to \infty.$$

8. Apply the Fibonacci or golden section searches to find the maximum of the functions defined in problems 1 or 2. Use the brackets already obtained and reduce the interval by a factor of 10.
9. Use the simplex technique of section 2.3.2 on the function

$$f = x^2 - 2xy + 2y^2;$$

continue until all the rules have been employed.

10. Apply the simplex method to the minimization of the function

$$g = x^2 + x^4y^2 + z^2.$$

11. Apply the alternating variable method of section 2.3.3 to the functions f in problem 9 and g in problem 10.
12. By using the Gram–Schmidt process, find three orthonormal vectors from

$$\mathbf{p}_1 = (2, 0, 0), \qquad \mathbf{p}_2 = (1, 1, 1), \qquad \mathbf{p}_3 = (2, 0, 1).$$

13. Show how Rosenbrock's method works, up to the completion of the first change of axes, on f in problem 9. Start at (1, 1) with steps in the coordinate directions of 0.2, 0.2.
14. Given the function $f(x_1, \ldots, x_n)$, show that the vector ∇f calculated at a point P is perpendicular to the contour, $f =$ constant, through P.
15. Evaluate the sequence of points starting at (0, 0) defined by the steepest descent method for the function

$$F = (x - y)^2 + (y - 1)^2.$$

Estimate the improvement in the convergence if Booth's modification is used; this method only proceeds to 0.9 of the distance to the minimum for each search direction.
16. Verify that equation (2.9) holds for the Taylor expansion of a three variable function.
17. Use Newton's method, section 2.4.2, (i) on the quadratic function F in problem 15, (ii) on the function g in problem 10.
18. Decide whether the matrices

$$\begin{bmatrix} 1 & 0 \\ 1 & 1 \end{bmatrix}, \quad \begin{bmatrix} 1 & 0 \\ 1 & -1 \end{bmatrix}, \quad \begin{bmatrix} 1 & 0 & 1 \\ 0 & 1 & 0 \\ -1 & 0 & 1 \end{bmatrix}$$

are positive definite or not.
19. Apply the Davidon procedure, with updating methods (2.20), (2.21), (2.22) on the quadratic function f in problem 9 and on g in problem 10.

Chapter III

Hill Climbing with Constraints

3.1 General Comments

It was noted in chapter 1 that constraints are extremely important in optimization work and if operative they usually dominate the solution. The most important and certainly the best used technique in constrained optimization concerns *linear programming*. Although it is not really classed as a hill climbing technique it is included in this chapter (section 3.4) as perhaps the most appropriate place. The method deals with the optimization of a *linear* function subject to *linear* constraints. This linearity gives the problem a particularly simple form and a vast amount of interesting mathematics can be performed with such a system. Full justice cannot be done to the array of basic theorems available or to the sophisticated range of computational algorithms, but some attempt will be made in section 3.4 to explain the problem and the simplex method used to solve the problem.

If, however, either the function or the constraints are *non-linear*, the difficulty of the problem can increase enormously and hill climbing techniques are required. For equality constraints, as in section 1.2.3, it is possible to change the problem to an unconstrained one either by using Lagrange multipliers or by solving each of the constraints for one of the variables. As a numerical procedure, Lagrange multipliers increase the number of variables and tend to bring in a large number of saddle points; both of these tend to make the convergence of the unconstrained method more difficult. Solving for a constraint can be a time-consuming operation but, if possible, because of the simplicity of the equation or perhaps some other prior knowledge, it is helpful to the numerical method since it eliminates the constraint completely and reduces the number of unknowns. Either of these two approaches can be useful, however, since the problem is reduced to an unconstrained problem considered in chapter 2. The only extensive experience of the author in constrained optimization has been with constraints which are all linear. Even here, where it might be thought that elimination would be reasonably easy, when the number of constraints is around fifteen to twenty it becomes easier *not* to eliminate but to use a modified version of the Davidon technique, see Fletcher (1968), which works directly on the constraints. In this particular work, it was found desirable to use Lagrange multipliers on two of the constraints since both these multipliers could be shown to have physical significance, they were easily estimated and they were required in the final calculation anyway. In general, however, if constraints can be avoided by elimination, transformation or some such device, this is usually preferable.

Inequality constraints pose a more formidable problem as can be seen from figure 3.1. It may be noted that if the maximum is required in the region $x \geq 0$ then it is at A in the *interior* of the region at a local and global

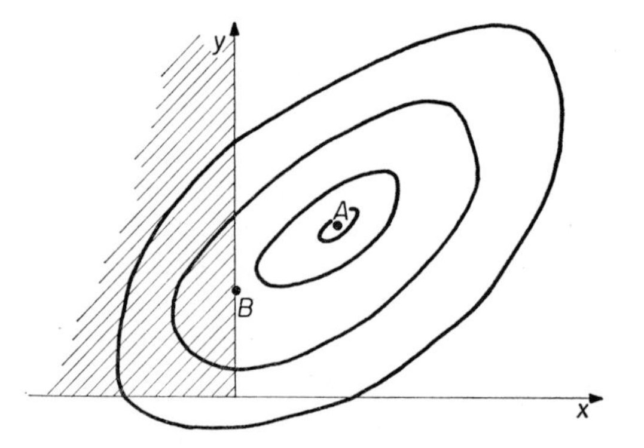

Figure 3.1 Contours of a constrained function.

maximum; if on the other hand the maximum is required in the region $x \leq 0$ then it is at B on the *boundary* of the region at a global but not a local maximum. Here lies the main difficulty of such work since the search must cover both the interior and the boundary. It may be recalled that in section 1.2.1 similar comments were made when considering the one dimensional problem illustrated in figure 1.1. In the author's problem mentioned in the above paragraph, inequality constraints of the type $0 \leq x \leq 1$ were operative on all the variables but it was known that in almost all cases the solution was interior to the region; hence if a constraint was violated then the search was returned to the interior of the feasible region. Finding this feasible region can be a non-trivial matter, particularly for the initial point, but there are standard computational methods of doing this.

The main aim of this chapter is to illustrate one or two of the available constrained optimization algorithms by the use of simple examples. In general, constrained methods require much more technical detail than unconstrained methods and complete implementations are usually long and complex. Such detail will not be given here but various articles in Lootsma (1972) discuss these points and give reference to where many methods and programmes may be found. One particular class of methods that has recently gained in popularity but is not discussed in this chapter uses Lagrange multipliers. The methods rely on necessary conditions, derived by Kuhn and Tucker (1951), for the extremum of a function subject to inequality constraints. They are the analogue of the Lagrange multiplier Result III in section 1.2. The conditions on the derivatives of the function, the constraint function and on the Lagrange multipliers have been incorporated into an efficient computational algorithm. Fletcher (1972) is a useful source for the detail of these methods. Besides being of use in such algorithms, the Kuhn–Tucker conditions lead to some extremely interesting mathematics. This is one possible direction that a mathematically inclined reader could usefully pursue (see, for example, Aoki (1971)).

There are adaptations of many standard unconstrained methods to deal with inequality constraints, for instance one possible method of solution is to use the random search procedure mentioned in section 2.1. The only additional rule that must be added is that if any of the constraints are violated, the calculation is stopped and the next random point is chosen. The method still has the considerable merit of being very easy to programme, but it has the usual disadvantage of being inefficient. Other methods can similarly be adapted to deal with constraints. With derivatives available, it has already been mentioned that Davidon's technique can be modified to deal with linear constraints; however, the details are very technical and will not be considered. For problems with no derivative available, Rosenbrock (1960) considered the necessary modifications that allow his technique to deal with constraints. Basically the method records a failure if a step violates the constraints at any time but otherwise the method proceeds in the usual way. A little more care has to be taken to deal satisfactorily with the boundary regions and the reader is referred to the original article for details. The simplex technique of section 2.3.2 can be adapted similarly, recording a failure if a new point lies outside the feasible region. It is found that using only regular simplices is not very satisfactory and the Nelder and Mead (1965) idea of using irregular simplices provides much better results, particularly when the optimum is on the boundary (see Box (1966)).

Before proceeding to consider methods in any detail two examples will be quoted to illustrate the advantages and disadvantages of eliminating equality constraints.

Example 23 Find the minimum of the function $(x + y)$ subject to

$$\left(\frac{a}{x}\right)^2 + \left(\frac{b}{y}\right)^2 = 1.$$

This can clearly be done using a Lagrange multiplier or by a straight elimination, but a more straightforward method is to put

$$\frac{a}{x} = \cos\theta, \qquad \frac{b}{y} = \sin\theta,$$

when the constraint is satisfied identically and the optimum is now required of

$$(a \sec\theta + b \operatorname{cosec}\theta).$$

The problem can be set up naturally in this way if the geometrical situation is considered. A ladder is to be taken horizontally round a 90° bend in a corridor, with width a in one arm, b in the other. This is illustrated in figure 3.2. Standard calculus, Result I, gives the minimum at $\theta = \tan^{-1}[(b/a)^{\frac{1}{3}}]$

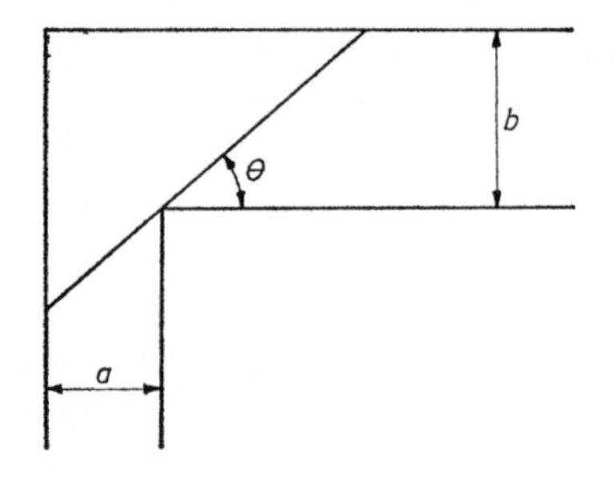

Figure 3.2 Ladder round a 90° bend.

and illustrates the benefit of eliminating a constraint. For a more complicated problem the minimization would almost certainly need a hill climbing procedure.

Example 24 Find the minimum of

$$y = \frac{x^{\frac{1}{2}}}{1 + x}. \tag{3.1}$$

Although no constraint is written explicitly in this problem, for the function to be real it is necessary that $x \geq 0$. An obvious way to overcome this constraint is to write $x = z^2$ thus ensuring that x is non-negative. Now (3.1) becomes

$$y = \frac{z}{1 + z^2} \tag{3.2}$$

and

$$\frac{dy}{dz} = \frac{1 - z^2}{(1 + z^2)^2}$$

with a minimum at $z = -1$ which is also the global minimum of (3.2). The actual minimum of (3.1) is not the local minimum given by $z = -1$, and hence $x = 1$, but is at $x = 0$. The paradox is explained by inverting $x = z^2$ to give $z = \pm(x)^{\frac{1}{2}}$ and noting that (3.1) and (3.2) are not the same function unless the correct sign is chosen at all times. An unconstrained hill climbing technique on (3.2) would have given $z = -1$ with the contradictory result of a minimum at $x = 1$. A great deal of computational effort would be wasted in this way.

Perhaps the warning of this example is that if any substitution is made, it is necessary to think through the possibility of producing spurious answers of the above type before launching into computation. As in this example for constrained problems the global extremum is usually required, which is not necessarily a local extremum. This presents the major computational difficulty of preventing a method converging to a spurious local extremum. Often the best that can be done is to start the calculation at several feasible points. If convergence to the same point is obtained it is highly probable that the global extremum has been achieved.

3.2 Gradient Projection

In methods which select a direction for a single variable search, for instance steepest descent or Davidon, a useful modification involves gradient projection. The chosen method is followed until one of the single variable searches violates one or more constraints. The 'best' point in this current search direction is then taken as the point *on* the boundary. The new direction of search is computed; if this sends the search point into the feasible region, the search is continued as normal; if, however, the constraints are immediately violated then the new direction is projected on to the tangent planes to the violated constraints. This projected direction is then followed to select the optimum in the feasible region. If no such point can be found an alternative procedure must select a direction back into the feasible region, preferably enhancing the function as well.

As an example find the maximum of

$$f = x^2 + y^2$$

subject to

$$x \geq 0, \qquad y \geq 0, \qquad (x + 2y - 1) \leq 0, \qquad 4 - (4x + 1)(2y + 1) \geq 0$$

using the steepest ascent algorithm, section 2.4.2. The region defined by these constraints is shown in figure 3.3. The reader will soon convince himself by looking at the contours of f that the point A (0.75, 0) gives the optimum with objective function $f = 9/16$.

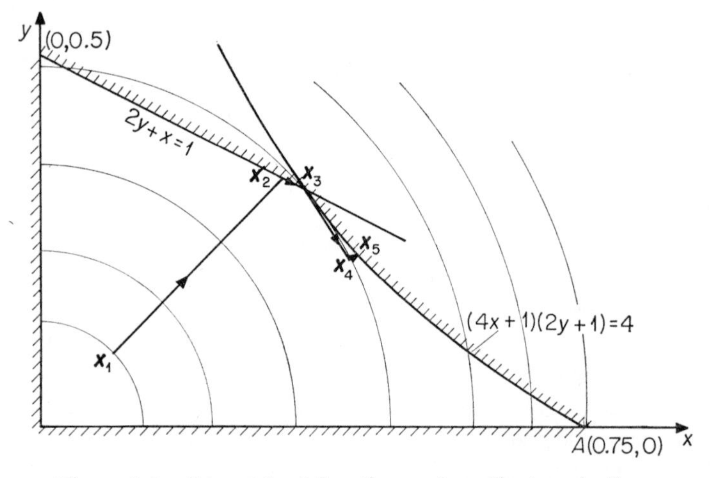

Figure 3.3 Constrained function and gradient projection.

The gradient of f is $\nabla f = (2x, 2y)$ so that if the search is started at $\mathbf{x}_1 = (0.1, 0.1)$, $(\nabla f)_1 = (0.2, 0.2)$. The direction of search is therefore

$$x = 0.1 + \lambda 0.2, \qquad y = 0.1 + \lambda 0.2,$$

and the objective function becomes

$$F(\lambda) = f(0.1 + \lambda 0.2, 0.1 + \lambda 0.2) = 0.02(1 + 2\lambda)^2.$$

This function clearly increases without bound as λ increases and achieves its maximum in the feasible region at the boundary point $\mathbf{x}_2 = (\frac{1}{3}, \frac{1}{3})$. The gradient at this point is $(\nabla f)_2 = (\frac{2}{3}, \frac{2}{3})$ and any movement in this direction takes the test point out of the feasible region. This direction is then projected down on to the tangent to the operative constraint $x + 2y - 1 = 0$. In a two dimensional situation such as this one, the projected direction is just the tangent direction itself, but in higher dimensions the detailed projection formulae must be worked out (see problem 4). The tangent direction to this straight line $x + 2y - 1 = 0$ is just the line itself and the new direction of search is therefore

$$x = \tfrac{1}{3} + 2\lambda, \qquad y = \tfrac{1}{3} - \lambda$$

with objective function

$$G(\lambda) = f(\tfrac{1}{3} + 2\lambda, \tfrac{1}{3} - \lambda) = (\tfrac{1}{3} + 2\lambda)^2 + (\tfrac{1}{3} - \lambda)^2.$$

Again G increases without bound as λ increases, so that the maximum value of λ, still in the feasible region, is at the corner B or $\mathbf{x}_3 = (0.359\,6, 0.320\,2)$. The gradient at $\mathbf{x}_3$ again takes the test point out of the feasible region so it is necessary to project once more. At B two constraints are operative; going through the formal procedure of projection produces the result that in a two dimensional problem it is impossible to project along two tangents at the same time. It is necessary therefore to decide how to proceed; a useful way of doing this is to omit the oldest operative constraint and then project. In the present case this certainly works and the projection is required along the tangent to the curve $(4x + 1)(2y + 1) = 4$. This is shown in figure 3.3, the maximum in this direction occurring at $\mathbf{x}_4 = (0.384\,9, 0.286\,2)$. Now none of the constraints are operative, so a normal steepest ascent step is taken; looking at the figure again the optimum is on the boundary at $\mathbf{x}_5$. The method then proceeds until the optimum is obtained.

The above is a simplified version of a gradient projection technique and it avoids various difficulties. (1) Because of the simplicity of the functions chosen the single variable searches could be done explicitly but in general they will need the much slower and less certain hill climbing methods of section 2.3, 2.4. (2) Even when a direction is projected on to the tangent planes of non-linear constraints it frequently happens that this direction also leads outside the feasible region. (3) When a corner point is reached it may happen that *any* direction at this point shows no improvement in the function, leaving the question of whether the optimum has been reached or whether the global optimum is elsewhere.

In both cases (2) and (3) some rules must be constructed to overcome these difficulties. It is often debatable how to proceed and the prescriptions are usually quite complex. It is therefore left to more exhaustive books (e.g. Beveridge and Schechter (1970)) or original papers for further details. Suffice it to add here that the basic idea of the method has been given and if used together with an efficient ascent method the process can be very effective.

3.3 PENALTY FUNCTIONS

This method is perhaps a more obvious way of dealing with constraints. Suppose a minimum of the function f is required subject to the constraint

$g \geq 0$, it seems natural to consider the minimization of

$$F = f + \frac{\lambda}{g}, \qquad \lambda > 0.$$

If the boundary of the constraint, $g = 0$, is approached then the function F will become large and positive and it can be assumed that any minimum seeking method will keep well away from this boundary. The difficulty, however, is that the minimum frequently occurs *on* the boundary; this has been stressed throughout the chapter. It is necessary therefore to devise a scheme, based on the above idea, that will also allow a boundary point to be selected. Carroll (1961) produced such a method which proceeds roughly as follows. Choose a value of λ (not too big or the constraint will dominate; not too small or extremely steep gradients will be encountered) and find the minimum of F by any suitable unconstrained method. If the minimum lies outside the feasible region then make the effect of the boundary more dominant by multiplying λ by 10. The process is repeated until a point, $\mathbf{x}_1$ say, is found within the feasible region. Starting at this point a new unconstrained search is commenced with λ replaced by $\lambda/10$, giving a minimum at $\mathbf{x}_2$. The λ is again divided by 10 and a new search initiated from the point $\mathbf{x}_2$. This process is repeated so that a sequence of minima $\mathbf{x}_1, \mathbf{x}_2, \mathbf{x}_3, \ldots$ is obtained, eventually tending to the global minimum in the region as $\lambda \to 0$. The method is usually called the *created response surface* technique and it can be applied equally well to many constraints. The problem has been reduced to a sequence of unconstrained searches and when used together with a Davidon technique, for instance, a powerful procedure in constrained optimization is obtained. Some care must be taken, however, since the convergence of the method is not assured and when things go wrong remedial action must be taken. Consideration of these points and advice on detailed applications, particularly in choosing an initial λ, can be found in Fiacco and McCormick (1964, 1966).

Example 25 Use the created response surface technique to find the minimum of

$$f = \frac{1}{1 + x} \qquad \text{with } x \leq 1.$$

The function F is written

$$F = \frac{1}{1 + x} + \frac{\lambda}{1 - x},$$

and it is plotted for several values of λ in figure 3.4. For this simple single variable example the optimization at each stage can be performed by classical calculus techniques, but normally an unconstrained hill climbing method would be required.

Choosing $\lambda = 1$ gives the minimum value as $x_1 = 0$, $F = 2$; now putting $\lambda = 0.1$ gives $x_2 = 0.520$, $F = 0.866$; with $\lambda = 0.01$, $x_3 = 0.818$, $F = 0.605$; similarly with $\lambda = 0.001$, $x_4 = 0.939$, $F = 0.516$, etc. It is clear from figure 3.4 (see also problem 5) that these values will tend eventually to the global minimum in the region at $x_{\min} = 1.000$ and $f = 0.500$. Two further points should be noted, firstly that for small values of λ the functions have very steep gradients and secondly that the previous solution gives a reasonable first guess for the new minimum.

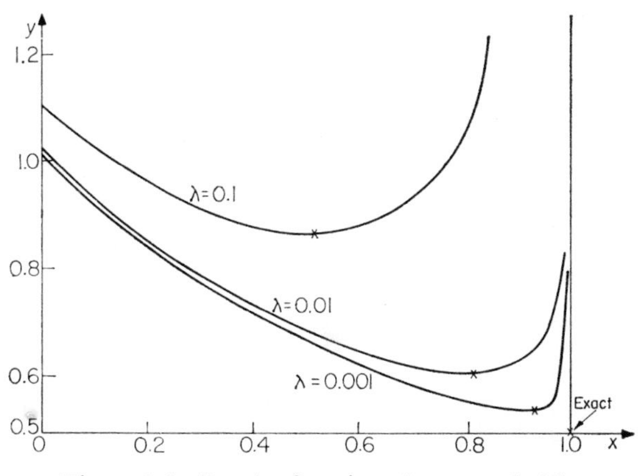

Figure 3.4 Penalty functions for example 25.

3.4 Linear Programming

Consider now the extremely important and very well used linear programming situation. A typical problem of this type is the following.

Find the maximum of

$$f = ax + by + cz$$

with

$$\alpha x + \beta y + \gamma z \leq \delta$$

$$\alpha' x + \beta' y + \gamma' z \leq \delta'$$

and

$$x \geq 0, \qquad y \geq 0, \qquad z \geq 0.$$

These problems assume that profits, costs, percentages, etc all behave linearly and in many situations this can be taken to be a reasonable assumption. It is perfectly feasible to deal successfully with problems of the above type containing several hundred variables. Before describing the basic simplex method, devised originally by Danzig in the early 1940s, consider the geometrical nature of a specific two variable problem. Find the maximum of

$$f = x + y$$

subject to

$$0 \leq x \leq 1, \qquad 0 \leq y \leq 1, \qquad 3x + 4y \leq 6.$$

The region under consideration is illustrated in figure 3.5 together with the contours of f = constant. It may be noted from these contours of f that the largest possible value of f that intersects the region is at A with $x = 1, y = \frac{3}{4}$, and $f = \frac{7}{4}$. A little thought leads to the conclusion that the optimum value

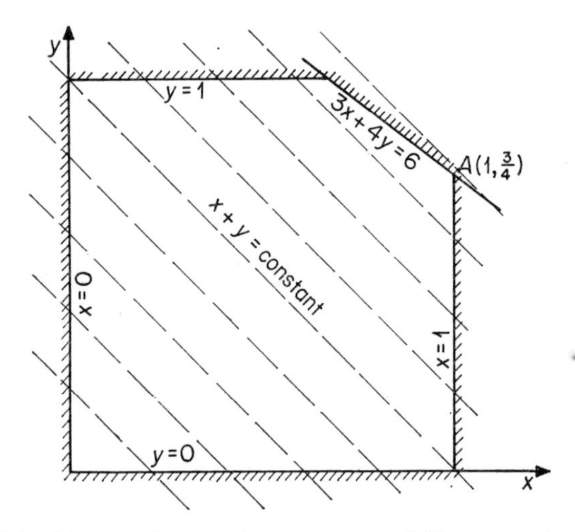

Figure 3.5 Linear programming, contours of f in constrained region.

will always be at a 'corner' of the region. Exceptions can occur; for instance if the f contours are parallel to a constraint then it is possible to have *every* point on this constraint as the optimum; problems can also be constructed that have *no* solution or *no finite* solution. However, for the present purposes concentrate on problems that have a unique finite solution.

The geometric intuition of the above simple problem applies quite generally to many dimensional problems, the detailed theorems and conditions establishing the result are well documented and to be found in any standard textbook on linear programming (for example, Danzig (1963) or Beveridge and Schechter (1970)).

The whole problem has now been considerably simplified since instead of having to look at *every* point in the feasible region it is now necessary only to look at the 'corners'. Even though the problem is now a finite one, the number of 'corners' can become very large so an efficient method of searching them must be employed. The simplex method starts at a particular vertex and systematically searches through the neighbouring vertices until one is found that increases the value of the objective function. This procedure is then repeated at the new vertex and continued until a vertex is found such that all the neighbours show no increase in the objective function. Provided the region under consideration is convex, it can be shown (and it is eminently reasonable) that this vertex will be the optimum and can be obtained in a finite number of steps. In this context a convex region R implies that *any* two points of R are joined by a straight line lying entirely in R, as illustrated in figure 3.6.

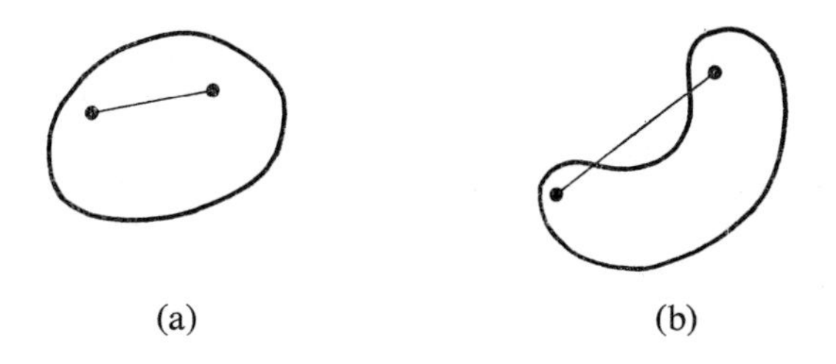

Figure 3.6 Convex region: (a) convex, (b) non-convex.

A computational algorithm is required to implement this procedure in practice. The rules of the algorithm will be developed for a particular example; they will then be stated in general without proof.

From a stock of 15, 20, 13 weight units of raw materials A, B and C it is required to manufacture products P, Q, R; P contains by weight 10% of A, 30% B, 60% C; Q contains 20% A, 30% B, 50% C; R contains 60% A, 40% B, and no C. The profits from P, Q, R are £2, £4, £6 per unit weight respectively. It is required to maximize the profit from the existing raw materials. Suppose that the amounts of P, Q, R produced are x, y, z then it is required to maximize

$$f = 2x + 4y + 6z \tag{3.3}$$

subject to

$$\begin{aligned} 0.1x + 0.2y + 0.6z &\leq 15 \\ 0.3x + 0.3y + 0.4z &\leq 20 \\ 0.6x + 0.5y \qquad &\leq 13 \\ x, y, z &\geq 0. \end{aligned}$$

It is usual at this stage to introduce *slack* variables u, v, w (first multiplying the above inequalities by 10) by

$$x + 2y + 6z + u = 150 \tag{3.4}$$

$$3x + 3y + 4z + v = 200 \tag{3.5}$$

$$6x + 5y + w = 130 \tag{3.6}$$

$$x, y, z, u, v, w \geq 0,$$

thus making all the inequalities of the same sort. Choose as a first guess at the solution $x = y = z = 0$ (these zero variables are usually called the non-basic variables) and $u = 150$, $v = 200$, $w = 130$ (these non-zero variables are called the basic variables); the function f in (3.3) takes the value 0. To move to a neighbouring vertex it may be decided to increase z keeping $x = y = 0$ and varying u, v, w accordingly; in any move of this type f will certainly increase. From (3.4) z can be increased to 150/6 while u decreases to zero; from (3.5) z can be increased to 200/4 while v decreases to zero. If the latter were chosen the value $z = 50$ would make $u < 0$ from (3.4) and one of the constraints is violated. Thus choose the smaller of the two values $z = 25$, giving $x = y = u = 0$, $v = 100$ from (3.5), $w = 130$ from (3.6) and finally $f = 150$. A clear improvement has therefore been made. Change the problem back to one of the above type, with non-basic variables x, y, u and basic variables v, w, z. Writing f in terms of the non-basic variables can be achieved by eliminating z between (3.4), (3.3)

$$f = 150 + x + 2y - u. \tag{3.7}$$

Similarly (3.4), (3.5), (3.6) can be rewritten so that the basic variables occur, one in each equation

$$\frac{1}{6}x + \frac{1}{6}u + \frac{1}{3}y + z = 25 \tag{3.8}$$

$$\frac{7}{3}x - 4u + \frac{5}{3}y + v = 200 \tag{3.9}$$

$$6x + 5y + w = 130. \tag{3.10}$$

Now choose to increase y, keeping $x = u = 0$ and varying the z, v, w accordingly. From (3.8), (3.9), (3.10) y can be increased to 75, 120, 26 respectively. The least of these is 26 so use this value of y; the solution then becomes $y = 26$, $w = x = u = 0$, $z = 49/3$, $v = 470/3$ and $f = 202$.

Again writing (3.7) in terms of w, x, u gives

$$f = 202 - \frac{7}{5}x - u - \frac{2}{5}w.$$

Increasing any of x, u, w from their zero values cannot increase f so the optimum value has been found; this geometrically implies that all the neighbouring points have smaller values. This optimum with solution $x = 0$, $y = 26$, $z = 49/3$ implies a maximum profit of £202, with all the raw materials A and C used while only $14\frac{1}{3}$ units of raw material B have been used.

For this particular problem the method is seen to work satisfactorily, but to make any real progress the method must be automated to avoid some of the apparently haphazard work above. In a hand computation this is best performed in a tableau display and will be illustrated for a three variable problem with three constraints. The method is quite general, however, and all the detailed steps in the manipulation can be established.

Table 3.1 Linear programming, simplex method: (a) illustration, (b) general tableau.

		x	y	z
u	150	1	2	6*
v	200	3	3	4
w	130	6	5	0
f	0	−2	−4	−6

(a)

		x_4	x_5	x_6
x_1	v_1	α_{14}	α_{15}	α_{16}
x_2	v_2	α_{24}	α_{25}	α_{26}
x_3	v_3	α_{34}	α_{35}	α_{36}
f	F	β_4	β_5	β_6

(b)

In the tableau the basic variables are written on the left with their values and the function and its value. Across the top are the non-basic variables (i.e. those with zero value). The rest of the array is made up of the coefficients in (3.4), (3.5), (3.6) and bottom line consists of the negative of the coefficients in (3.9). The rules of the procedure, which will not be proved, are as follows:

(1) Select the column with the smallest β value, say β_j: (-6).
(2) Evaluate v_1/α_{1j}, v_2/α_{2j}, v_3/α_{3j} and select the minimum of these, say v_i/α_{ij}: (150/6 marked *).
(3) Interchange x_i, x_j: (u, z, in (a)).
(4) Treat the first column and bottom row just as part of the matrix, say as column α_{l0} and row α_{0k}.
(5) Replace α_{ij} by $1/\alpha_{ij}$.
(6) Replace α_{il} by α_{il}/α_{ij} (all l with $l \neq j$).
(7) Replace α_{kj} by $-\alpha_{kj}/\alpha_{ij}$ (all k with $k \neq i$).
(8) Replace α_{kl} by $\alpha_{kl} - (\alpha_{kj}\alpha_{il}/\alpha_{ij})$ (all k, l with $k \neq i$, $l \neq j$).

The particular array in table 3.1(a) then reads

Table 3.2 Linear programming, tableau after first application of the simplex method.

		x	y	u
z	50	1/6	1/3	1/6
v	100	7/3	5/3	−2/3
w	130	6	5*	0
f	150	−1	−2	1

following the above rules. It will be noted that the entries are identical with the coefficients in (3.7) . . . (3.10). This process is then repeated again and the rules give

Table 3.3 Linear programming, final tableau.

		x	w	u
z	49/3	−7/30	−1/3	1/6
v	470/3	1/3	−5/3	−2/3
y	26	6/5	1/5	0
f	202	7/5	2/5	1

The process is now terminated since all the entries in the bottom row (or β's) are positive and no improvement in f can be achieved.

This tableau procedure can be followed for any number of variables and any number of constraints and various computer implementations of this method are available. It was pointed out earlier in the geometrical discussion that for various reasons the method can fail. Equally in this analytic discussion the same failures can occur; these must, of course, be taken care of in a full discussion. For the present purposes, however, the basic problem and the basic solution procedure have been demonstrated and it is left to the interested reader to consult the quoted references for a discussion of the different cases.

PROBLEMS

1. By using a suitable parameterization find the minimum of

$$x^2 + y^2$$

subject to

$$x^{\frac{2}{3}} + y^{\frac{2}{3}} = 1.$$

2. Eliminate the constraint in the function

$$y = x^2[(x^2 - 1)^{\frac{1}{2}}], \qquad |x| \geq 1$$

and find the minimum of this function.

3. Use the gradient projection method of section 3.2 to maximize

$$f = 2x^2 + y^2$$

subject to the constraints $x \geq 0$, $\frac{1}{2} \geq y \geq 0$, $x + y \leq 1$.

4. If the direction $\nabla f(x_1, \ldots, x_n)$ is projected down on to the plane

$$\sum_{i=1}^{n} a_i x_i = 1,$$

show that this projected direction is parallel to

$$[\mathbf{p}^2\, \nabla f - (\mathbf{p} \cdot \nabla f)\mathbf{p}],$$

where $\mathbf{p} = (a_1, \ldots, a_n)$.

5. In example 25 show that for a given $\lambda > 0$ the minimum of $F = (1 + x)^{-1} + \lambda(1 - x)^{-1}$ is given by $\bar{x} = [1 - (\lambda)^{\frac{1}{2}}]/[1 + (\lambda)^{\frac{1}{2}}]$. Thus show that as $\lambda \to 0$, $\bar{x} \to 1$ and $\bar{F} \to \frac{1}{2}$.

6. Use the penalty function method for minimizing the constrained function given in problem 3. Note that several penalty functions must be used, but only use one multiplying parameter.

7. Solve the linear programming problem of maximizing the function

$$5x + 4y + 6z$$

subject to

$$\begin{aligned} 4x + y + z &\leq 20 \\ 3x + 4y + 6z &\leq 30 \\ 2x + 4y + z &\leq 25 \\ x + y + 2z &\leq 15 \end{aligned}$$

and $x, y, z \geq 0$.

8. A manufacturer produces three types of cloth in various combinations of colours. One yard of cloth uses amounts of coloured wool as follows:

Table 3.4 Cloth composition, problem 8.

	Cloth type *A*	*B*	*C*	Amount available
green	1	2	1	10
red	2	1	2	6
blue	3	1	0	10

From one yard length of cloth the profit from types *A*, *B*, *C* are 3, 5, 4 units respectively. If the wool is to be used to maximize the total profit (i) find how much profit is made, (ii) determine how much wool is left unused.

Chapter IV

Heuristic and Formal Approach to Variational Methods

4.1 Heuristic Approach

Consider first one of the variational examples which has an obvious geometric solution and was described in chapter 1.

Example 6 Find the shortest distance between two points (0, 0) and (a, 0) in a plane.

It was shown in section 1.3 that this problem could be reduced to finding the minimum length, s, from

$$s = \int_0^a (1 + y'^2)^{\frac{1}{2}}\, dx, \qquad y(0) = y(a) = 0 \tag{4.1}$$

where $y = y(x)$ has a continuous first derivative. It is clear geometrically that the curve $y = 0$ gives the extremum. This can be proved rigorously since

$$(1 + y'^2)^{\frac{1}{2}} \geq 1 \qquad \text{for all } y$$

and hence

$$\int_0^a (1 + y'^2)^{\frac{1}{2}}\, dx \geq \int_0^a 1\, dx = a.$$

Since there is only one function, $y = 0$, which satisfies the boundary conditions and gives equality in the above it follows that this is the required solution.

Suppose now that this *geodesic* problem is made a little more difficult by asking for the shortest distance between two points on a cone. Geometrically a construction is still possible since a paper cone can be snipped down a generator, laid out flat, the two points joined by a straight line and then the cone pasted together again. This job sounds a difficult one to write analytically but it is not in fact impossible (see problem 2). If, however, the problem is further complicated by asking for the shortest distance between two points on a general surface it is quite clear that the heuristic approach described has to be formalized into a precise mathematical procedure.

The following is an example with a less obvious physical or geometrical interpretation.

Example 26 Find the twice differentiable function $u(x)$ which makes

$$\int_0^1 u'^2 \, dx \qquad \text{with } u(0) = 1,\ u(1) = 0 \tag{4.2}$$

a minimum.

Here one must proceed by inspired guesswork, which suggests (at least to the inspired) that a quadratic function should be tried. Indeed a little more careful thought would suggest a rather more general function. However, concentrate first on the quadratic

$$u = a + bx + cx^2. \tag{4.3}$$

To satisfy the boundary conditions: $u(0) = 1$ gives $a = 1$ and $u(1) = 0$ gives $b + c + 1 = 0$. Hence (4.3) takes the form

$$u = (1 - x)(1 + Ax). \tag{4.4}$$

(Note that the more careful thought might have suggested

$$u = (1 - x)(1 + Ax + Bx^2 + Cx^3 + \cdots)) \tag{4.5}$$

Putting (4.4) into (4.2) gives

$$\int_0^1 u'^2 \, dx = \int_0^1 [(A - 1)^2 - 4A(A - 1)x + 4A^2x^2] \, dx,$$

which is just a function of A and when the integrations are performed

$$\int_0^1 u'^2 \, dx = f(A) = 1 + \tfrac{1}{3}A^2.$$

Minimizing this function with respect to A

$$0 = \frac{df}{dA} = \frac{2}{3}A.$$

Hence out of the class of functions (4.4) the one which makes (4.2) a minimum is

$$u = (1 - x) \quad \text{and} \quad \int_0^1 u'^2 \, dx = 1.$$

The next question to ask is whether a *better* result can be obtained. Try therefore the general function

$$u = (1 - x) + g(x) \qquad \text{with } g(0) = g(1) = 0. \tag{4.6}$$

The boundary conditions are satisfied and it is just a matter of substituting this form into (4.2),

$$\begin{aligned}\int_0^1 u'^2\,dx &= \int_0^1 (-1 + g')^2\,dx = \int_0^1 (1 - 2g' + g'^2)\,dx \\ &= 1 - 2[g]_0^1 + \int_0^1 g'^2\,dx \\ &= 1 + \int_0^1 g'^2\,dx \geq 1.\end{aligned}$$

The integrated term goes to zero because of the boundary conditions. It is only when $g' = 0$ and hence $g = 0$ (to satisfy the boundary conditions) that equality is obtained and the function $u = (1 - x)$ is therefore the best available.

This example should be noted very carefully and understood in detail since (4.5) *contains the germ of the idea behind approximate methods and* (4.6) *contains the germ of the idea behind the formal Euler equation approach.*

Consider a further example which this time has some physical interest.

Example 27 A uniform thin beam of length l is clamped at its ends and is loaded continuously by a constant force F/unit length. Find the displacement of the beam.

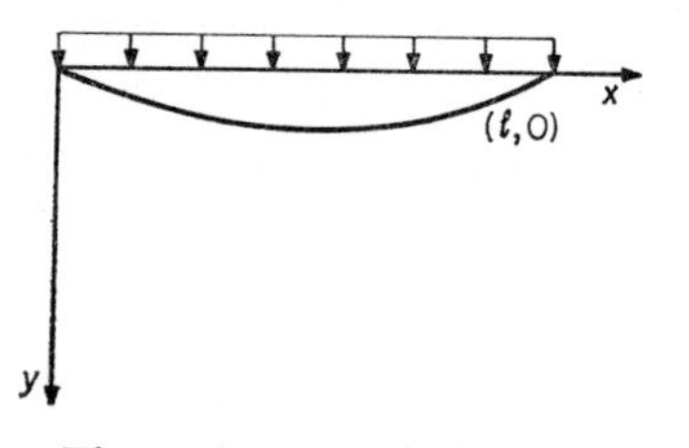

Figure 4.1 Loaded beam.

The configuration is illustrated in figure 4.1. It will be shown in chapter 9 that the total energy V of the system for small displacements is

$$V = \int_0^l [\tfrac{1}{2}(EI)y''^2 - Fy]\,dx, \tag{4.7}$$

where y is the displacement and (EI) is a constant. The clamping conditions at the ends give

$$y(0) = y'(0) = y(l) = y'(l) = 0.$$

This problem corresponds to a simple model of a bridge under its own weight, the final configuration of the bridge being obtained by minimizing the total energy over all y.

Consider a polynomial function for y, satisfying the four end conditions and leaving one free parameter

$$y = Ax^2(x - l)^2. \tag{4.8}$$

Substituting this form into (4.7) gives

$$V(A) = \tfrac{1}{2}(EI)A^2 \int_0^l (12x^2 - 12xl + 2l^2)^2\, dx - FA \int_0^l x^2(x - l)^2\, dx$$

$$= \left[\frac{1}{2}(EI)\frac{4}{5}l^5\right] A^2 - \left(\frac{Fl^5}{30}\right) A.$$

Minimizing with respect to A gives

$$A = \frac{F}{24EI},$$

giving in turn the best result using (4.8) as

$$y = Y = \frac{Fx^2(x - l)^2}{24EI}. \tag{4.9}$$

Again the problem is to try to establish whether this form (4.9) is the best possible result or just the best result that can be derived from (4.8). The method is again to extend (4.9) to the trial function

$$y = Y + g,$$

where Y is given by (4.9) and g is an unknown function which ensures that the boundary conditions are satisfied or

$$g(0) = g'(0) = g(l) = g'(l) = 0.$$

Putting $\alpha = F/EI$ and substituting this form into (4.7) gives

$$\frac{V}{EI} = \int_0^l (\tfrac{1}{2}Y''^2 + Y''g'' + \tfrac{1}{2}g''^2 - \alpha Y - \alpha g)\, dx. \tag{4.10}$$

Concentrating for the moment only on the second term in the integral, this can be integrated by parts to give

$$\int_0^l Y''g''\,dx = [Y''g']_0^l - \int_0^l Y'''g'\,dx$$

$$= -[Y'''g]_0^l + \int_0^l Y''''g\,dx$$

$$= \int_0^l Y''''g\,dx,$$

where the integrated terms are put equal to zero because of the boundary conditions. The terms involving Y alone in (4.10) can be integrated since Y is a known function and (4.10) becomes

$$\frac{V}{EI} = -\frac{\alpha^2 l^5}{1\,440} + \int_0^l g(Y'''' - \alpha)\,dx + \int_0^l \tfrac{1}{2}g''^2\,dx.$$

Since $Y'''' = \alpha$ from the definition (4.9), the middle term in this expression disappears and as the final term is always positive the form Y gives the exact minimum for V.

Several interesting and very pertinent questions can be asked about this deduction. For instance, is the fact that $Y'''' = \alpha$ fortuitous or was it to be expected? What would happen if the function (4.9) were not the exact minimum? If a different form had been assumed for V would the same technique be useful? Can necessary and sufficient conditions be always found for such problems? All these questions need careful consideration and it is clear that the problem must be written down generally and the methods used above formalized into precise mathematical procedures.

It is inevitable that when new problems such as those described in this section are encountered, a heuristic approach must be used. Eventually, however, once a body of knowledge and experience has been built up it becomes tedious to perform the same work for each new problem and it is essential to look at common characteristics and then to try to produce a general theory. Once this has been done the next step is to push the theory to the limit of its applicability and then beyond when again one is thrown back on to a heuristic approach.

4.2 Some Definitions Concerning Functionals

It must be stressed again that when forms such as (4.1), (4.2), (4.7) are studied, for each y or u selected from a given class of functions a *number* is evaluated. The problem is then to find the function which gives the greatest or least number as the function varies through the whole class chosen. This idea of a number being associated with a function is just the idea used to define a *functional.*

Definition A *functional* is an operator which maps a given set on to the real numbers. (There are complex functionals which map to the complex numbers but these will not be considered.) The functionals of interest in this work are essentially mappings from some class of functions on to the real line, see figure 4.2, and are usually defined by integrals.

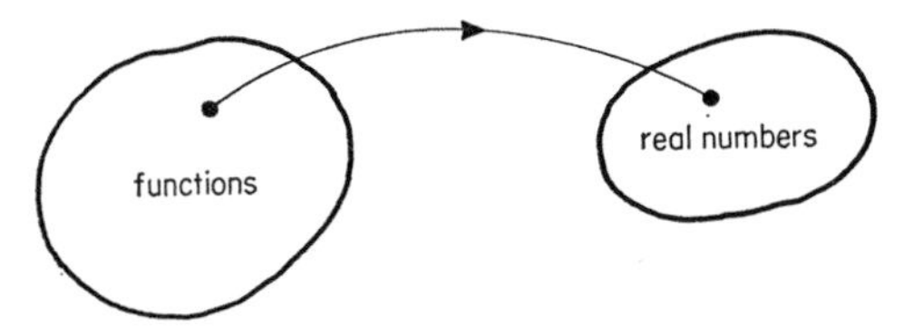

Figure 4.2 Functional as a mapping.

Example 28

(*a*) $\int_0^a (1 + y'^2)^{\frac{1}{2}}\, dx$ with $y(0) = y(a) = 0$;

(*b*) $\iint_R (\nabla\varphi)^2\, dx\, dy$ with φ given on the boundary of R.

These two examples illustrate the type of functional encountered in this work. In (a) y must be selected from the class of functions for which the integrals exists, for instance functions with a continuous first derivative on $[0, a]$. In (b) choose $\varphi(x, y)$ from the class of functions defined on R and with continuous first partial derivatives.

The *notation* used in this book for a functional is to write the mappings $f[y]$. This then reminds the reader that y is *not* a number but a function. In more sophisticated treatments this notation is unnecessary since functionals are placed in a more abstract setting and results can be specialized to numbers or functions as appropriate. Indeed the whole of the calculus of variations can be treated abstractly as a branch of functional analysis. The results obtained in the next two chapters then differ very little from the results obtained in chapter 1. This, however, requires careful definitions of continuity and differentiability of mappings and a long detailed look at Banach spaces. This approach is not particularly suitable for a book at this level. It is, however, an extremely important approach and is being increasingly used by workers in optimization, Luenberger (1969). The simpler approach used here comes across several serious difficulties which can only be sorted out satisfactorily by further abstraction. A typical example concerns the limits of sequences of functions (see example 31). Unless the space of functions is chosen carefully it is possible to construct sequences with limits which do not belong to the space. The completion of the space is an important concept but it requires a considerable effort to sort it out satisfactorily.

Having stressed the importance of functional analysis and an abstract approach, which will be touched on again in chapter 13, the present approach deals almost entirely with functionals of the form

$$F[y] = \int_a^b f(x, y, y', y'', \ldots)\, dx$$

$$G[\varphi] = \iint_R g(x, y, \varphi, \varphi_x, \varphi_y, \ldots)\, dx\, dy$$

$$H[y, z] = \int_a^b h(t, y, z, \dot{y}, \dot{z}, \ldots)\, dt \tag{4.11}$$

together with obvious extensions of these. One word of explanation of the interpretation for instance of $f(x, y, y')$. Here the function f is written as if x, y, y' were independent variables of an ordinary function of three variables. It will be found necessary to evaluate $\partial f/\partial y$, $\partial f/\partial y'$ etc and these derivatives should be interpreted in this way. In example 6 for instance

$$f(x, y, y') = (1 + y'^2)^{\frac{1}{2}}, \qquad \frac{\partial f}{\partial y'} = y'(1 + y'^2)^{-\frac{1}{2}}.$$

This notation is not a very satisfactory one but it is standard and will be used throughout.

A further piece of notation that will be extensively used is to let $C^n[a, b]$ denote the class of functions with continuous nth derivatives in $[a, b]$. Where there is no ambiguity over the interval this set is just written C^n.

The position has now been reached when a fairly general problem can be identified, one which certainly includes most of the examples so far discussed.

Major problem *If $u \in C^2[a, b]$ and*

$$u(a) = A, \qquad u(b) = B,$$

where A and B are given numbers, find the function which makes the functional

$$J[u] = \int_a^b f(x, u, u')\, dx$$

an extremum.

Once this problem has been solved a whole variety of similar problems (different functionals, several variables, several functions, different boundary conditions, etc) follow by an almost identical method. The basic approaches have been set in example 26 and it is just a matter of rewriting the method for the general problem cited above. The first of these methods is to use an approximation procedure, with suitable trial functions chosen, thus reducing the problem to a standard optimization of a function of many variables. This line of attack is followed in chapter 5. The second approach is to assume that the solution is known and then to attempt to prove that the assumption was correct or at least to find some conditions on the assumed solution. This leads to the Euler equation which is dealt with in detail in chapter 6.

PROBLEMS

1. With the parametric representation of a sphere

$$x = a \sin \theta \cos \varphi, \qquad y = a \sin \theta \sin \varphi, \qquad z = a \cos \theta,$$

construct a formula for the length of a curve $\varphi = \varphi(\theta)$ between points $(0, 0, a)$, $(a \sin \theta_0, 0, a \cos \theta_0)$. Hence deduce from first principles that a geodesic on a sphere is a great circle.

2. Repeat, with a suitable parameterization, the above procedure for the cone

$$x^2 + y^2 = z^2 \tan^2 \alpha.$$

3. Show that the functional

$$\int_0^1 (y'^2 + y^2)\, dx, \qquad y(0) = 0, \qquad y(1) = 1,$$

has its extremum for $Y = \sinh x / \sinh 1$. Consider the function $y = Y + g$ with $g(0) = g(1) = 0$.

4. Use a quintic in x as a trial for the extremum of the beam functional (4.7) with $F = kx$ and boundary conditions $y(0) = y'(0) = y(l) = y'(l) = 0$.

5. Given the set $S = \{x : 0 < x < 1\}$, show that for the sequence $a_n = (1/n) \in S$, $\lim_{n \to \infty} a_n \notin S$.

6. Given the sequence of functions $y = s_n(x)$, $y > 0$ defined by the hyperbolas $y^2 = x^2 + (1/n^2)$, show that as $n \to \infty$, $s_n \to |x|$. Thus although s_n are differentiable everywhere, at $x = 0$ the limit function is not differentiable.

Chapter V

Approximate Methods in Variational Problems

5.1 Trial Functions

This method has been anticipated in examples 26 and 27 of the previous chapter where polynomial approximations were chosen as trial functions. Both examples, however, were chosen so that the exact solution could be guessed easily. It was indicated that a suitable sequence of trial functions could be written down and used even if the solution could not be spotted. First reconsider example 26.

Example 26 Find the function that minimizes

$$J[u] = \int_0^1 u'^2\, dx \qquad \text{with } u(0) = 1,\ u(1) = 0.$$

The exact solution was computed as $U = (1 - x)$ with $J[U] = 1$. Now suppose a trigonometric trial function is chosen to satisfy the boundary conditions, $u_1 = \cos \frac{1}{2}\pi x$, then evaluating $J[u_1] = \pi^2/8 = 1.234$. A second trial can be attempted again satisfying the end conditions

$$u_2 = \cos \tfrac{1}{2}\pi x + A \sin \pi x$$

$$J[u_2] = \int_0^1 (-\tfrac{1}{2}\pi \sin \tfrac{1}{2}\pi x + A\pi \cos \pi x)_2\, dx$$

$$= \tfrac{1}{8}\pi^2 + \tfrac{2}{3}\pi A + \tfrac{1}{2}A^2\pi^2.$$

Minimizing this function of A gives

$$0 = \frac{2}{3}\pi + A\pi^2 \quad \text{or} \quad A = -\frac{2}{3\pi}.$$

Hence

$$U_2 = \cos\frac{1}{2}\pi x - \frac{2}{3\pi}\sin \pi x, \qquad J[U_2] = 1.012.$$

A third trial

$$u_3 = \cos \tfrac{1}{2}\pi x + \alpha \sin \pi x + \beta \sin 2\pi x$$

gives

$$\alpha = -\frac{2}{3\pi}, \qquad \beta = -\frac{1}{15\pi} \quad \text{and} \quad J[U_3] = 1.003.$$

The following table gives the numerical values derived from the exact solution and the three trial approximations.

Table 5.1 Exact and approximate solutions of example 26.

x	0	0.2	0.4	0.6	0.8	1.0	$J[u]$
$1 - x$	1	0.8	0.6	0.4	0.2	0	1
u_1	1	0.951 1	0.809 0	0.587 8	0.309 0	0	1.234
u_2	1	0.826 4	0.607 2	0.386 0	0.184 3	0	1.012
u_3	1	0.806 2	0.594 7	0.398 5	0.204 5	0	1.003

It may be noted that the accuracy of the trial functions increases rapidly as the number of parameters increases. In fact the maximum error in u_2 is about 8% and in J about 1%, while for u_3 the maximum error is about 2% and in J about 0.3%. This looks a promising approach since it appears that good accuracy can be achieved at the expense of a few tedious but trivial integrations.

The following is an example which has not been tackled before and which can be attacked in the same manner.

Example 29 Find the extremum of the functional

$$J[y] = \int_{-\frac{1}{2}}^{+\frac{1}{2}} [(1 - x^2)y'^2 - 2y^2]\, dx$$

where

$$y(-\tfrac{1}{2}) = 0 \quad \text{and} \quad y(\tfrac{1}{2}) = 1.$$

This problem can be solved exactly in terms of the Legendre functions $P_1(x) = x$ and $Q_1(x) = \frac{1}{2}x \log [(1 + x)/(1 - x)] - 1$ as

$$y = x + \frac{\frac{1}{2}Q_1(x)}{Q_1(\frac{1}{2})}.$$

It is a non-trivial exercise to obtain this solution, see chapter 6, and any slight complication would immediately make an exact solution quite impracticable.

A first trial function just satisfying the boundary conditions is

$$y_1 = x + \tfrac{1}{2}, \qquad J[y_1] = \tfrac{1}{4}.$$

This can obviously be improved by writing

$$y_2 = x + \tfrac{1}{2} + A(\tfrac{1}{4} - x^2),$$

the boundary conditions are still satisfied and the parameter A can be chosen to minimize J. Since this form includes y_1 as a special case it certainly cannot do worse.

$$J[y_2] = F(A) = \int_{-\frac{1}{2}}^{+\frac{1}{2}} \{(1 - x^2)(1 - 2Ax)^2 - 2[x + \tfrac{1}{2} + A(\tfrac{1}{2} - x^2)]^2\}\, dx$$

$$F(A) = \frac{1}{4} - \frac{1}{3}A + \frac{13}{60}A^2.$$

Minimizing with respect to A

$$F'(A) = 0 = -\frac{1}{3} + \frac{13}{30}A$$

and hence

$$Y_2 = x + \frac{1}{2} + \frac{10}{13}\left(\frac{1}{4} - x^2\right), \qquad J[Y_2] = 0.122.$$

The approximation can be improved still further by writing

$$y_3 = x + \tfrac{1}{2} + (\tfrac{1}{4} - x^2)(\alpha + \beta x),$$

a form which satisfies the boundary conditions and includes y_2 as a special case. When substituted into the functional, a function of the two variables α and β is obtained which, when the partial derivatives are put to zero, gives the next approximation. In fact it is found that $\alpha = 10/13$, $\beta = 0$ so that no improvement is obtained. This could have been anticipated from the properties of $Q_1(x)$ which insists that in the exact solution any addition to the $(x + \frac{1}{2})$ term must be an *even* function of x. The further approximations are trivial in principle but need considerable effort in the integrations. A comparison of these first two approximations is given in the table 5.2.

Table 5.2 Exact and approximate solutions of example 29.

x	−0.5	−0.3	0	0.3	0.5	$J[y]$
y_{exact}	0	0.325 3	0.689 3	0.925 3	1	0.121
Y_1	0	0.2	0.5	0.8	1	0.250
Y_2	0	0.323 1	0.692 3	0.923 1	1	0.122

For more difficult problems where the exact solution is not known it may be hoped that the good accuracy obtained in these two examples would follow. This in fact is often the case but the very good accuracy of these two examples is obtained because these solutions are very well behaved and examples can be constructed when serious difficulties occur (see problem 2).

A general expansion for the solution of this last example can be written

$$y_N = (x + \tfrac{1}{2}) + (\tfrac{1}{4} - x^2)(\alpha_0 + \alpha_1 x^2 + \cdots + \alpha_N x^{2N}).$$

Suppose the minimum of J, selecting from functions of this form, is obtained at $y = Y_N$ then several questions can be asked. For instance does $J[Y_N] \to$ exact minimum as $N \to \infty$ or does Y_N tend in some sense the function giving the exact minimum? The questions are not easy to answer but some attempt will be made at them in the next section.

The method used in the above is not, of course, restricted to problems involving a single variable and many variable problems can be attacked in a precisely similar way.

Example 30 Find the minimum of the functional

$$J[T] = \iint_R (\nabla T)^2 \, dx \, dy,$$

where R is the region $|x| \leq 1, |y| \leq 1$ and $T = 0$ on $x = 1$, $T = (1 - x)$ on $y = \pm 1$ and $T = 2y^2$ on $x = -1$.

This problem corresponds physically to calculating the steady state temperature distribution of a square plate with the given temperatures on its edges. It will be seen later that the present calculation gives an approximate solution of the Laplace equation.

By symmetry the trial functions must depend on y^2 and to satisfy the boundary conditions T can be chosen as

$$T = (1 - x)y^2 + (1 - x^2)(1 - y^2)(A + Bx + Cy^2 + Dx^2 + Exy^2 + \cdots).$$

The first term satisfies the edge conditions while the remainder is zero round the whole boundary.

1st trial

$$T_1 = (1 - x)y^2, \qquad J[T_1] = (356/45) = 7.91.$$

2nd trial

$$T_2 = (1 - x)y^2 + A(1 - x^2)(1 - y^2)$$

$$\nabla T_2 = [-y^2 - 2Ax(1 - y^2), 2(1 - x)y - 2Ay(1 - x^2)]$$

$$J[T_2] = \int_{-1}^{+1}\int_{-1}^{+1} \{[y^2 + 2Ax(1 - y^2)]^2 + 4y^2[1 - x - A(1 - x^2)]^2\} \, dx \, dy$$

$$= \frac{4}{45}(89 - 80A + 64A^2).$$

Minimizing this function of the single variable A gives $A = \frac{5}{8}$ and $J[T_2] = (256/45) = 4.69$.

Further trials can be made, the main effort required being the detailed manipulation and integrations which are tedious and very susceptible to error. For problems of this sort no exact solution is known except from numerical solution of the Laplace equation. It is important therefore to be able to predict the accuracy of the approximations. This is a recurring difficulty and at some stage a serious effort must be made to try to sort it out. For comparison a numerical solution T_N, using a mesh size 0.25, was evaluated and these values and the T_2 values are presented in table 5.3 for sample points in the region.

Table 5.3 Comparison of the numerical solution T_N and T_2.

(x, y)	$(-\frac{1}{2}, 0)$	$(0, 0)$	$(\frac{1}{2}, 0)$	$(-\frac{1}{2}, \frac{1}{2})$	$(0, \frac{1}{2})$	$(\frac{1}{2}, \frac{1}{2})$
T_N	0.55	0.56	0.33	0.82	0.69	0.38
T_2	0.47	0.63	0.47	0.73	0.72	0.48

It is seen immediately for a two dimensional case the accuracy is nowhere near as good as the cases already studied and further approximations are absolutely essential.

5.2 Rayleigh–Ritz Technique

The ideas that have been developed in this and the last chapter are now formalized into a fairly generally applicable technique. Take the general problem quoted in section 4.2.

General problem Find the extremum of

$$J[u] = \int_a^b f(x, u, u')\, dx \tag{5.1}$$

with

$$u(a) = \alpha, \qquad u(b) = \beta; \qquad \alpha, \beta \text{ given.}$$

Choose sufficiently differentiable functions $u_0, u_1, u_2, \ldots$ so that

$$u_0(a) = \alpha, \qquad u_0(b) = \beta$$
$$u_i(a) = u_i(b) = 0, \qquad i = 1, 2, 3, \ldots.$$

Now consider the trial function

$$U_N = u_0(x) + \sum_{i=1}^{N} A_i u_i(x), \tag{5.2}$$

which clearly satisfies the boundary conditions of the problem. Substitute this U_N into (5.1) and perform all the integrations then

$$J[U_N] = G(A_1, A_2, \ldots, A_N).$$

Find the extremum of this function G which is just an ordinary function of the A_i and can be solved by Result II of chapter 1. Suppose this extremum is given by $A_i = \bar{A}_i$ then the best result for U_N is

$$\bar{U}_N = u_0 + \sum_{i=1}^{N} \bar{A}_i u_i(x).$$

This is just the *Rayleigh–Ritz* technique for the functional (5.1) and the same basic method generalizes to more complex situations, for instance in example 30.

One way of interpreting this method is to first note that the whole class, $\mathscr{C}$, of functions from which u is selected is an extremely large set. The next step is then to choose a more manageable set $\mathscr{C}_1 \subset \mathscr{C}$ (e.g. U_1 in (5.2)) and then find the extremum for this set. Having done this, select a larger set $\mathscr{C}_2$ (e.g. U_2 in (5.2)) with $\mathscr{C}_1 \subset \mathscr{C}_2 \subset \mathscr{C}$ and again find the extremum for this set. Building up sets in this way, a whole sequence of nested subsets is considered $\mathscr{C}_1 \subset \mathscr{C}_2 \subset \mathscr{C}_3 \subset \cdots \subset \mathscr{C}$ and for each of these subsets the extremum is found by the much simpler techniques described in Results I or II in chapter 1.

The method in either interpretation leaves basic questions to be answered:

(1) Will any sequence $\{u_i\}$ or $\{\mathscr{C}_i\}$ do?
(2) Does $J[\bar{U}_N] \to J[U_{\text{exact}}]$ as $N \to \infty$?
(3) In some sense does $\bar{U}_N \to U_{\text{exact}}$ as $N \to \infty$?

To question 1 the answer is clearly 'no' since it is necessary to choose $\{u_i\}$ which are capable of approximating any function in $\mathscr{C}$. Such a set is called a *complete set* of functions and it can be defined as follows. Given any $f \in \mathscr{C}$ then for any $\varepsilon > 0$ there exists N and numbers $\alpha_1, \alpha_2, \ldots, \alpha_N$ so that

$$|f - \alpha_1 u_1 - \alpha_2 u_2 \cdots - \alpha_N u_N| < \varepsilon$$

for all x in the range $[a, b]$. Such complete sets are not particularly easy to establish; the best known are the *polynomials* $1, x, x^2, \ldots$ which form a complete set for $C^0[a, b]$, the class of continuous functions on a finite interval $[a, b]$ and Fourier series or *trigonometric functions* 1, $\sin(\pi x/a)$, $\cos(\pi x/a)$, $\sin(2\pi x/a)$, $\cos(2\pi x/a)$, ... which form a complete set for $\mathscr{C}$ the class of continuous functions with period $2a$ over any interval of length $2a$. The study of complete sets is a subject in itself and although it is both interesting and essential for detailed proofs of convergence of the Rayleigh–Ritz technique, it is rarely considered in practical calculations. This is principally because it is unusual, even with advanced computing techniques, to evaluate more than the first few approximations and good physical or geometrical insight give a clear indication of the type of functions to use.

A partial answer to question (2) can be given quickly. Let $J[\bar{U}_N] = K_N$ and suppose a minimum is sought. The set $\mathscr{C}_{N+1}$, built up in (5.2), includes $\mathscr{C}_N$, i.e. $\mathscr{C}_{N+1} \supset \mathscr{C}_N$, and hence the worst that can happen is that no improvement is made or $K_{N+1} \leq K_N$. Thus the sequence $\{K_N\}$ is monotonic decreasing and either tends to a limit or $-\infty$. If $K = J[U_{\text{exact}}]$ exists then $K_N \geq K$ for all N and the sequence is therefore bounded below. Thus the sequence tends to a limit and the partial answer is that $J[\bar{U}_N]$ tends to a limit. Whether this limit is the correct limit $J[U_{\text{exact}}]$ is another matter which requires a detailed study. In fact it can be shown that provided $\{u_i\}$ form a complete set the correct limit is obtained, but it is left for the reader to consult a more advanced treatise for the detailed proof.

The third question asked is the most difficult to answer and only an indication of the complexities can be given. To establish the result needs very stringent conditions on the functional, the set of admissible functions and the approximating set. A very well quoted example is sufficient to illustrate the pitfalls of considering sequences of functions.

Example 31 Consider the sequence of functions $\{s_n(x)\}$

$$s_n(x) = \frac{1 + nx}{1 + n^2x^2} \qquad \text{as } n \to \infty.$$

These functions are very well behaved, being differentiable infinitely many times. It would appear that as $n \to \infty$, $s_n \to 0$ (i.e. the zero function) but putting $x = 1/n$ gives $s_n(1/n) = 1$. Consequently for *every* s_n there are points which take the value 1. In the limit $n \to \infty$, the limit function is one which is zero everywhere except at $x = 0$ where it takes the value unity. It is certainly not even continuous at the origin, a somewhat disturbing fact considering the $s_n(x)$ are differentiable infinitely many times.

This example is not untypical and readers familiar with Fourier analysis will recognize similar difficulties that occur near discontinuities (Gibbs' phenomenon). Before any serious attack can be made on answering questions such as question 3 a full study of classical and functional analysis is essential. It would be desirable for any given problem to prove that the approximating sequence tends to the exact answer but this task is normally far too big and it is just assumed to be the case.

Despite the warnings about sequences of functions it is interesting to consider the particularly simple case of a functional involving only quadratic terms. The following example illustrates the problem, the basic approach and method of solution.

Example 32 Apply a general Rayleigh–Ritz technique to the functional

$$J[u] = \int_a^b uu''\,dx,$$

with

$$u(a) = \alpha, \qquad u(b) = \beta, \qquad u \in C^2.$$

Let $u_0, u_1, \ldots$ be chosen so that $u_0(a) = \alpha$, $u_0(b) = \beta$, $u_i(a) = u_i(b) = 0$ for all $i \geq 1$ and $u_i \in C^2$. Now the test function for the extremum is

$$u = u_0 + \sum_{i=1}^{n} a_i u_i,$$

where it is clear that all the subsidiary conditions are satisfied. Substituting into the functional gives

$$J[u] = \int_a^b \left(u_0 u''_0 + \sum_{i=1}^{n} a_i(u_0 u''_i + u_i u''_0) + \sum_{i,j=1}^{n} a_i a_j u_i u''_j \right) dx.$$

Defining

$$L_{ij} = \int_a^b u_i u''_j\,dx \quad \text{and} \quad c_i = -\tfrac{1}{2}\int_a^b (u_0 u''_i + u_i u''_0)\,dx$$

the functional takes the form

$$J[u] = L_{00} - 2\sum_{i=1}^{n} a_i c_i + \sum_{i,j=1}^{n} L_{ij} a_i a_j.$$

It should be noted that the L's and c's are known numbers calculated by integrating suitable combinations of the chosen functions. The extremum of this functional is now given by

$$\frac{\partial J}{\partial a_i} = 0 = -2c_i + 2\sum_{j=1}^{n} L_{ij} a_j, \qquad i = 1, \ldots, n,$$

or rewriting in matrix form the equations become

$$\begin{bmatrix} L_{11} & \cdots & L_{1n} \\ \cdot & & \cdot \\ \cdot & & \cdot \\ \cdot & & \cdot \\ L_{n1} & \cdots & L_{nn} \end{bmatrix} \begin{bmatrix} a_1 \\ \cdot \\ \cdot \\ \cdot \\ a_n \end{bmatrix} = \begin{bmatrix} c_1 \\ \cdot \\ \cdot \\ \cdot \\ c_n \end{bmatrix}. \tag{5.3}$$

Thus the Rayleigh–Ritz technique for this problem produces a set of linear equations which can be solved by straightforward matrix inversion. The first and second approximations to a_1, for instance, would be

$$a_1 = \frac{c_1}{L_{11}}, \qquad a_1 = \frac{\begin{vmatrix} c_1 & L_{12} \\ c_2 & L_{22} \end{vmatrix}}{\begin{vmatrix} L_{11} & L_{12} \\ L_{21} & L_{22} \end{vmatrix}}.$$

The above analysis is applicable to a wide class of functionals which contain quadratic terms. It will be seen in the next chapter that optimization of such functionals corresponds to the solution of linear differential equations. It should be noted that the tremendous power of matrix algebra can now be used to get a computational solution of (5.3) and to decide on the existence of such solutions. It is instructive to apply a similar technique to other functionals and the reader is urged to try problem 8 at the end of this chapter.

5.3 Relation to Hill Climbing

In all the variational problems considered so far the problem has reduced to solving a set of linear equations or equivalently a matrix problem. This is primarily because the functionals chosen have all contained quadratic costs, i.e. terms of the form $u^2, uu', u'u'', \ldots$. This is not, of course, always the case and many problems have much more complicated functionals. Two examples are now considered to illustrate difficulties with the Rayleigh–Ritz method and the relation to the hill climbing techniques of chapter 2.

Example 33 Find the extremum of the functional

$$J[\varphi] = \int_0^1 (\varphi'^2 + \varphi^3)\, dx$$

with

$$\varphi(0) = 0, \qquad \varphi(1) = 1.$$

For trial functions an obvious sequence can be obtained by successively truncating the series

$$\varphi = x + x(1 - x)(A + Bx + Cx^2 + \cdots).$$

Consider the first approximation,

$$\varphi_1 = x + Ax(1 - x)$$

$$J[\varphi_1] = \int_0^1 [1 + A(1 - 2x)]^2 + [x + Ax(1 - x)]^3\, dx$$

$$= \frac{5}{4} + \frac{3}{20}A + \frac{23}{60}A^2 + \frac{1}{140}A^3.$$

The extremum of this function of A is easily obtained but it should be noted that the function is a cubic. The extrema are found at $A = -0.196\,7$ and -35.58, the first corresponding to a minimum and the second to a maximum.

The second approximation is

$$\varphi_2 = x + x(1 - x)(A + Bx)$$

$$J[\varphi_2] = \int_0^1 \{[1 + A(1 - 2x) + B(2x - 3x^2)]^2 + x^3[1 + (1 - x)(A + Bx)]^3\}\, dx.$$

Again all the integrals can be performed to give a function of the form

$$a_1 + a_2 A + a_3 B + a_4 A^2 + a_5 AB + a_6 B^2 + a_7 A^3 + a_8 A^2 B + a_9 AB^2 + a_{10} B^3,$$

where the a_i are known from the integrations. Finding the extremum of this function of A and B is not an easy task and reduces to the solution of two non-linear simultaneous equations in A and B. It was precisely at this stage that the hill climbing techniques of chapter 2 were introduced. In this case the Davidon procedure would work very well on the above function.

A well known physical problem that will be studied in more detail in section 7.1 concerns the curve of quickest descent. The problem has a known but by no means obvious solution and it will be seen that application of a Rayleigh–Ritz technique leads to a hill climbing problem more difficult than the last example.

Example 34 Axes Oxy are set up with Ox horizontal and Oy vertically down. A smooth wire joins the points (0, 0) and $A(a, b)$. A bead slides down the wire under gravity from rest at O to A and it is required to find the shape of the wire to minimize the time of descent.

The energy equation of the bead, when it has speed v and has descended a vertical distance y, is

$$\tfrac{1}{2}mv^2 = mgy.$$

But $v^2 = [1 + (dy/dx)^2](dx/dt)^2$ and hence substituting and performing one integration gives the time of descent T as

$$T = \frac{1}{(2g)^{\frac{1}{2}}} \int_0^a \left(\frac{1 + y'^2}{y}\right)^{\frac{1}{2}} dx,$$

with $y(0) = 0$, $y(a) = b$ to ensure that the wire passes through the points O and A. Firstly write this problem in a non-dimensional form putting

$$T^* = T\frac{(2g)^{\frac{1}{2}}}{a}, \quad X = \frac{x}{a}, \quad a^{-1}y\left(\frac{x}{a}\right) = Y(X), \quad B = \frac{b}{a}$$

then

$$T^*[Y] = \int_0^1 \left(\frac{1 + Y'^2}{Y}\right)^{\frac{1}{2}} dX.$$

Taking first approximations of trigonometric and polynomial types

$$Y_1 = B \sin \tfrac{1}{2}\pi X + \alpha \sin \pi X, \qquad Y_2 = BX + \beta X(1 - X)$$

and substituting into the functional gives

$$G(\alpha) = T^*[Y_1] = \int_0^1 \left(\frac{1 + (\frac{1}{2}\pi B \cos \frac{1}{2}\pi X + \pi\alpha \cos \pi X)^2}{B \cos \frac{1}{2}\pi X + \alpha \sin \pi X}\right)^{\frac{1}{2}} dX$$

$$H(\beta) = T^*[Y_2] = \int_0^1 \left(\frac{1 + [B + \beta(1 - 2X)]^2}{BX + \beta X(1 - X)}\right)^{\frac{1}{2}} dX.$$

In both these cases the integrals are virtually impossible to evaluate explicitly and the functions G and H are best regarded as defined by the integrals. It will be noted that this problem is considerably more difficult than the previous example since each function evaluation requires a numerical integration. If this is the magnitude of the problem it is absolutely essential that to minimize G or H the most economical use of function evaluations must be made. It is precisely because of difficulties of this sort that new methods have developed to cope with these rather awkward problems and it will be shown how to attack this problem from a dynamic programming method in chapter 10.

These two examples show clearly that for functionals which do *not* depend quadratically on the unknown function a hill climbing technique is necessary. Whether this approach is the best one is doubtful and it is in many cases more profitable to use an alternative method.

PROBLEMS

1. Construct suitable trigonometric and polynomial trial functions for the minimum of

$$J[y] = \int_0^{\frac{1}{2}\pi} (y'^2 - y^2)\, dx, \qquad y(0) = 0, \qquad y(\tfrac{1}{2}\pi) = 1.$$

Compute the first unknown coefficient for each trial and compare the results.

2. Try a similar analysis to example 29 for the minimization of

$$J[y] = \int_0^1 [(1 - x^2)y'^2 - 2y^2]\, dx$$

with $y(0) = 1$, $y(1) = 0$. Note that difficulties will eventually arise because of the infinite nature of $Q_1(x)$ at $x = 1$; in fact the problem has no solution.

3. Find an approximate solution for the extremum of the functional

$$J[y] = \int_1^2 x(y'^2 - y^2)\,dx, \qquad y(1) = 1,\ y(2) = 2.$$

4. Construct a sequence of trial functions for the functional

$$J[u] = \iint_R (\nabla u)^2\,dx\,dy$$

with $\{R: (x, y), |x| \leq a, |y| \leq a\}$ and $u = 0$ on the sides of the square $y = \pm a$, $x = -a$ and $u = (a^2 - y^2)$ on $x = a$. Compute the first unknown coefficient by the Rayleigh–Ritz technique.

5. Find approximations to the extremum of

$$\int_0^1 (y''^2 + y^2)\,dx$$

subject to $y(0) = y'(0) = y''(0) = 0$, $y(1) = 1$.

6. Look for polynomial solutions for the optimum of

$$J[r, \theta] = \int_0^T \left(\frac{1}{2}\dot{r}^2 + \frac{1}{2}r^2\dot{\theta}^2 + \frac{\mu}{r}\right) dt$$

with $r(0) = a$, $\dot{r}(0) = 0$, $\theta(0) = 0$, $\dot{\theta}(0) = h/a$.

7. Show that the method of trial functions for

$$\int_0^a y(1 + y'^2)^{\frac{1}{2}}\,dx, \qquad y(0) = \alpha,\ y(a) = \beta$$

leads to a hill climbing problem. Set up the problem in such a way that it is amenable to such a technique.

8. Use the method on p. 81 in examples 26 and 30 of the text. In example 26 use the specific trial functions $u_0 = \cos \frac{1}{2}\pi x$, $u_p = \sin p\pi x$, $p = 1, 2, \ldots$.

Chapter VI

Euler Equation

6.1 Single Function of One Variable

Approximate or heuristic methods have been shown to be reasonably satisfactory for several variational problems. For a more complete understanding, however, it is necessary to look, if possible, for exact solutions for these problems. In several of the quoted examples it has been stated that the optimization of a functional is equivalent to solving a differential equation, so now is an appropriate moment to look at the deduction of this basic differential equation, the *Euler equation*. This will first require the proof of a basic lemma. Later sections will require natural extensions of this lemma but these will not be proved.

Basic lemma Let $\eta \in C^2[a, b]$ and satisfy

$$\eta(a) = \eta(b) = 0.$$

If f is continuous in $a \leq x \leq b$ and

$$\int_a^b f(x)\eta(x)\, dx = 0$$

for *all* $\eta \in C^2$ then $f \equiv 0$.

Proof The proof is by contradiction. Suppose $f > 0$ for some point c in $[a, b]$. Since f is continuous then $f > 0$ in a neighbourhood of c say $[c_1, c_2]$ with $a < c_1 < c < c_2 < b$. Define the function

$$\eta_1(x) = \begin{cases} 0 & a \leq x \leq c_1 \\ (x - c_1)^4(x - c_2)^4 & c_1 \leq x \leq c_2 \\ 0 & c_2 \leq x \leq b, \end{cases}$$

then $\eta_1 \in C^2$ as can easily be verified. But

$$\int_a^b f\eta_1 \, dx = \int_{c_1}^{c_2} f(x - c_1)^4(x - c_2)^4 \, dx > 0$$

since f is positive in $[c_1, c_2]$. Hence a contradiction against the hypothesis of the lemma has been derived. A similar contradiction follows if it is assumed that $f < 0$ and hence $f \equiv 0$ as required.

It was noted in section 4.1 that example 26 contained all the essential ingredients of the deduction of the Euler equation and this example should be studied before proceeding.

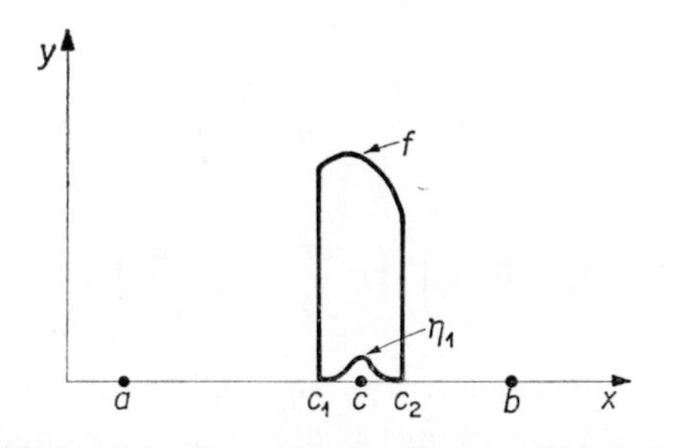

Figure 6.1 Test function for case $f(c) > 0$.

Consider now the general problem for a single function of one variable quoted in section 4.2; find the extremum of the functional

$$J[y] = \int_a^b f(x, y, y') \, dx,$$

with $y(a) = A$, $y(b) = B$, A, B given and y selected from the class C^2. It is too much to expect that sufficient conditions for the extremum can be deduced since this was a very non-trivial job for the simpler problem of optimizing a function of several variables. Necessary conditions, however, are much more straightforward and it is reasonable to assume that these can be found. Suppose therefore that $Y(x)$ is an extremum (a maximum say) of $J[y]$. Certainly to be an extremum, Y must satisfy the conditions of the problem that $Y \in C^2$ and $Y(a) = A$, $Y(b) = B$. To be a maximum of J then $J[Y] \geq J[y]$ for all y 'close to' Y. Generate these y by letting $\eta(x)$ be such that $\eta(a) = \eta(b) = 0$ and $\eta \in C^2$; then the function $y = Y + \varepsilon\eta$ satisfies

$$y(a) = A, \qquad y(b) = B, \qquad y \in C^2.$$

In the expression $y = Y + \varepsilon\eta$, ε is a number (small) and η is a function which is supposed known, although it will be seen later that the results deduced apply to all such functions. Substitute this function into the functional and expand by Taylor's theorem

$$J[Y + \varepsilon\eta] = \int_a^b f(x, Y + \varepsilon\eta, Y' + \varepsilon\eta')\, dx$$

$$= \int_a^b f(x, Y, Y')\, dx + \int_a^b \left(\varepsilon\eta \frac{\partial f}{\partial Y} + \varepsilon\eta' \frac{\partial f}{\partial Y'}\right) dx + 0(\varepsilon^2).$$

Since Y and η are known then all the integrals can be performed, at least theoretically, and the functional just becomes a function, $F(\varepsilon)$, of the single variable ε. It was assumed at the start that Y was an extremum and therefore $\varepsilon = 0$ is an extremum or $F'(0) = 0$. But

$$F'(\varepsilon) = \int_a^b \left(\eta \frac{\partial f}{\partial Y} + \eta' \frac{\partial f}{\partial Y'}\right) dx + 0(\varepsilon)$$

and hence

$$F'(0) = 0 = \int_a^b \left(\eta \frac{\partial f}{\partial Y} + \eta' \frac{\partial f}{\partial Y'}\right) dx.$$

The expression is almost ready for an application of basic lemma; an integration by parts and use of $\eta(a) = \eta(b) = 0$ is required first,

$$0 = \int_a^b \eta \left[\frac{\partial f}{\partial Y} - \frac{d}{dx}\left(\frac{\partial f}{\partial Y'}\right)\right] dx + \left[\eta \frac{\partial f}{\partial Y'}\right]_a^b$$

or

$$0 = \int_a^b \eta \left[\frac{\partial f}{\partial Y} - \frac{d}{dx}\left(\frac{\partial f}{\partial Y'}\right)\right] dx.$$

Since this same result can be obtained for any $\eta \in C^2$ satisfying $\eta(a) = \eta(b) = 0$, and assuming the bracketed term is continuous the basic lemma can be used to give

$$\frac{\partial f}{\partial Y} - \frac{d}{dx}\left(\frac{\partial f}{\partial Y'}\right) = 0.$$

One point in the above proof requires further comment. On expansion of f by Taylor's theorem it was tacitly assumed that $\varepsilon\eta$ and $\varepsilon\eta'$ are both small. Even if 'y is close to Y', so that $\varepsilon\eta$ is small there is no guarantee that $\varepsilon\eta'$ is small even though $\eta \in C^2$. If there is no restriction on the derivative of the approximating functions then the extremum is called a *strong* extremum but if a restriction is imposed that 'y' is also close to Y'' then the extremum is called *weak*. Thus by implication the Euler equation is a necessary condition for a weak extremum and in the rest of this book the word extremum in this context refers to the weak type. Because of the size of the class of functions involved, conditions for strong extrema are much more difficult to obtain and require a knowledge of functional analysis to make much progress.

RESULT IV

A necessary condition for the extremum of

$$J[y] = \int_a^b f(x, y, y')\, dx,$$

where $y(a) = A$, $y(b) = B$, *(A, B known) and* $y \in C^2$, *the class of functions with continuous second derivatives, is that y satisfies the Euler equation*

$$\frac{d}{dx}\left(\frac{\partial f}{\partial y'}\right) - \frac{\partial f}{\partial y} = 0.$$

Example 35 Find the Euler equations of

$$\text{(1)} \quad I[u] = \int_0^1 u'^2\, dx, \qquad u(0) = 1,\ u(1) = 0,$$

$$\text{(2)} \quad J[y] = \int_{-\frac{1}{2}}^{\frac{1}{2}} \{(1 - x^2)y'^2 - 2y^2\}\, dx, \qquad y(-\tfrac{1}{2}) = 0,\ y(\tfrac{1}{2}) = 1,$$

$$\text{(3)} \quad K[\varphi] = \int_0^1 (\varphi'^2 + \varphi^3)\, dx, \qquad \varphi(0) = 0,\ \varphi(1) = 1.$$

(1) $f(x, u, u') = u'^2$, $\partial f/\partial u = 0$, $\partial f/\partial u' = 2u'$ and hence the Euler equation is

$$0 = \frac{d}{dx}\left(\frac{\partial f}{\partial u'}\right) - \frac{\partial f}{\partial u} = \frac{d}{dx}(2u')$$

or

$$u'' = 0.$$

It is not surprising that in sections 4.1 and 5.1 solutions for this extremum were easy to guess.

(2) $g(x, y, y') = (1 - x^2)y'^2 - 2y^2$, $\partial g/\partial y = -4y$, $\partial g/\partial y' = 2(1 - x^2)y'$ and hence the Euler equation is

$$0 = \frac{d}{dx}[2(1 - x^2)y'] - (-4y)$$

or

$$(1 - x^2)y'' - 2xy' + 2y = 0,$$

which is just the Legendre equation of order 1.

(3) $h(x, \varphi, \varphi') = \varphi'^2 + \varphi^3$ and the Euler equation is

$$0 = \frac{d}{dx}(2\varphi') - 3\varphi^2$$

or

$$\varphi'' = \frac{3}{2}\varphi^2.$$

This is a non-linear second order differential equation and it is not unexpected that in example 33 some difficulty was encountered in the solution.

Special case An important special case of functions of the type being studied lead to an immediate first integral of the Euler equation. This special case involves functionals of the form

$$\int_a^b f(y, y')\, dx,$$

that is $f(y, y')$ has *no explicit* dependence on x. Observing that

$$df = \frac{\partial f}{\partial y} dy + \frac{\partial f}{\partial y'} dy'$$

or

$$\frac{df}{dx} = \frac{\partial f}{\partial y} y' + \frac{\partial f}{\partial y'} y''$$

the first integral of the Euler equation can be deduced from:

$$\frac{d}{dx}\left(y' \frac{\partial f}{\partial y'} - f\right) = y'' \frac{\partial f}{\partial y'} + y' \frac{d}{dx}\left(\frac{\partial f}{\partial y'}\right) - \frac{\partial f}{\partial y} y' - \frac{\partial f}{\partial y'} y''$$

$$= y' \left[\frac{d}{dx}\left(\frac{\partial f}{\partial y'}\right) - \frac{\partial f}{\partial y}\right] = 0.$$

This is put to zero since the extremum satisfies the Euler equation. An alternative way of looking at this result is to consider y' as an integrating factor of the Euler equation. The desired first integral is

$$y' \frac{\partial f}{\partial y'} - f = C. \tag{6.1}$$

Example 36 Find the first integrals of the Euler equations of

(1) $J[y] = \int_0^a (1 + y'^2)^{\frac{1}{2}}\, dx$ (cf. example 6)

(2) $I[\varphi] = \int_0^1 (\varphi'^2 + \varphi^3)\, dx$ (cf. example 33).

(1) $f(y, y') = (1 + y'^2)^{\frac{1}{2}}$ and the first integral is

$$C = y' \frac{y'}{(1 + y'^2)^{\frac{1}{2}}} - (1 + y'^2)^{\frac{1}{2}}$$

or

$$C(1 + y'^2)^{\frac{1}{2}} = y'^2 - 1 - y'^2 = -1.$$

Equivalently this can be written

$$y' = K$$

and hence the straight line solution for example 6.

(2) $g(\varphi, \varphi') = \varphi'^2 + \varphi^3$ and

$$C = \varphi'(2\varphi') - (\varphi'^2 + \varphi^3)$$

or

$$C = \varphi'^2 - \varphi^3.$$

This is the required first integral, which is now a non-linear first order differential equation that can be integrated once more

$$x = \int \frac{d\varphi}{(C + \varphi^3)^{\frac{1}{2}}} + K,$$

a result which can be evaluated in terms of elliptic integrals.

6.2 Several Functions of One Variable

The first generalization of Result IV is to functionals of the form

$$J[y_1, y_2, \ldots, y_n] = \int_a^b f(x, y_1, y_2, \ldots, y_n, y'_1, \ldots, y'_n)\, dx \tag{6.2}$$

with

$$y_i(a) = A_i, \qquad y_i(b) = B_i, \quad i = 1, 2, \ldots, n$$

and $y_i \in C^2$. The necessary condition for an extremum is obtained in the same manner as in section 6.1 with a small amendment at the end. Consider the two function case

$$J[y_1, y_2] = \int_a^b f(x, y_1, y_2, y'_1, y'_2)\, dx \tag{6.3}$$

$$\begin{aligned} y_1, y_2 \in C^2, \qquad y_1(a) = A_1, \qquad y_1(b) = B_1,& \\ y_2(a) = A_2, \qquad y_2(b) = B_2.& \end{aligned} \tag{6.4}$$

Let Y_1, Y_2 give the extremum and hence satisfy the conditions (6.4). For variations around these functions consider $\eta_1, \eta_2 \in C^2$, $\eta_i(a) = \eta_i(b) = 0$, $i = 1, 2$. The functions

$$y_1 = Y_1 + \varepsilon\eta_1, \qquad y_2 = Y_2 + \varepsilon\eta_2$$

therefore satisfy (6.4) and when substituted into (6.3) and the function f expanded, lead to

$$J[y_1, y_2] = \int_a^b f(x, Y_1, Y_2, Y'_1, Y'_2)\,dx$$
$$+ \varepsilon \int_a^b \left(\frac{\partial f}{\partial Y_1}\eta_1 + \frac{\partial f}{\partial Y'_1}\eta'_1 + \frac{\partial f}{\partial Y_2}\eta_2 + \frac{\partial f}{\partial Y'_2}\eta'_2 \right) dx + 0(\varepsilon^2).$$

Since Y_1, Y_2, η_1, η_2 are known functions this is just a function, $F(\varepsilon)$, of the single variable ε with an extremum at $\varepsilon = 0$. Hence

$$F'(0) = 0 = \int_a^b \left(\frac{\partial f}{\partial Y_1}\eta_1 + \frac{\partial f}{\partial Y'_1}\eta' + \frac{\partial f}{\partial Y_2}\eta_2 + \frac{\partial f}{\partial Y'_2}\eta'_2 \right) dx;$$

integrating by parts and using the fact that $\eta_i = 0$ at the end points

$$0 = \int_a^b \left\{ \eta_1 \left[\frac{\partial f}{\partial Y_1} - \frac{d}{dx}\left(\frac{\partial f}{\partial Y'_1}\right)\right] + \eta_2 \left[\frac{\partial f}{\partial Y_2} - \frac{d}{dx}\left(\frac{\partial f}{\partial Y'_2}\right)\right]\right\} dx. \tag{6.5}$$

The amendment to the previous argument is now required. The statement (6.5) is true for *all* known η_1, η_2 satisfying the appropriate conditions; in particular it is true for the special case $\eta_2 \equiv 0$. Thus applying the basic lemma to the first term of (6.5) gives

$$\frac{\partial f}{\partial Y_1} - \frac{d}{dx}\left(\frac{\partial f}{\partial Y'_1}\right) = 0.$$

Likewise the basic lemma can be applied to the second term of (6.5) with $\eta_2 \not\equiv 0$ and it similarly gives

$$\frac{\partial f}{\partial Y_2} - \frac{d}{dx}\left(\frac{\partial f}{\partial Y'_2}\right) = 0.$$

For the more general problem (6.2) this result clearly generalizes to the following: *a necessary condition for an extremum of* (6.2) *is that*

$$\frac{d}{dx}\left(\frac{\partial f}{\partial y'_i}\right) - \frac{\partial f}{\partial y_i} = 0, \qquad i = 1, 2, \ldots, n. \tag{6.6}$$

Example 37 Find the Euler equations of

$$\int_0^1 (yz - 2y^2 + y'^2 - z'^2)\,dx$$

$$0 = \frac{d}{dx}\left(\frac{\partial f}{\partial y'}\right) - \frac{\partial f}{\partial y} = \frac{d}{dx}(2y') - (z - 4y)$$

$$0 = \frac{d}{dx}\left(\frac{\partial f}{\partial z'}\right) - \frac{\partial f}{\partial z} = \frac{d}{dx}(-2z') - y$$

or the pair of equations

$$y'' = \tfrac{1}{2}z - 2y, \qquad z'' = -\tfrac{1}{2}y.$$

As a second example return to the geodesic problem left in section 4.1.

Example 38 Find the differential equations satisfied by the geodesic on a cone.

Let the cone be $x^2 + y^2 = z^2 \tan^2 \alpha$ with the parametric representation

$$x = z \tan \alpha \cos \theta, \qquad y = z \tan \alpha \sin \theta.$$

Any curve on the surface of the cone can be written $z = z(t)$, $\theta = \theta(t)$ and the length of the curve between t_1 and t_2 computed from

$$\dot{s}^2 = \dot{x}^2 + \dot{y}^2 + \dot{z}^2 = \sec^2 \alpha(\dot{z}^2 + \sin^2 \alpha z^2 \dot{\theta}^2). \tag{6.7}$$

Hence

$$s[z, \theta] = \sec \alpha \int_{t_1}^{t_2} (\dot{z}^2 + \sin^2 \alpha z^2 \dot{\theta}^2)^{\frac{1}{2}}\,dt.$$

Applying (6.6) for the extremum

$$0 = \frac{d}{dt}\left(\frac{\dot{z}}{\dot{s}}\right) - \frac{\sin^2 \alpha z \dot{\theta}^2}{\dot{s}}$$

$$0 = \frac{d}{dt}\left(\frac{\sin^2 \alpha z^2 \dot{\theta}}{\dot{s}}\right),$$

where $\dot{s}$ is given by (6.7). Replacing the independent variable t by the arc length s simplifies the equations considerably to

$$0 = z'' - \sin^2 \alpha z \theta'^2$$

$$c = z^2 \theta',$$

where dash denotes differentiation with respect to s. There still remains quite a difficult non-linear differential equation to solve before an explicit form for the geodesic is obtained. This basic method is applicable to a general geodesic problem, where the arc length is determined by $ds^2 = \Sigma g_{ij} dq_i dq_j$ and it will be looked at in more detail in section 7.4.

The form of the Euler equations (6.6) is principally used in classical mechanics where they relate very closely to the usual Lagrange equations, see section 7.3.

6.3* Multiple Integrals

A further generalization, studied satisfactorily by approximate techniques, involves multiple integrals; for instance example 30 involved the functional

$$\iint (\nabla\varphi)^2 \, dx \, dy.$$

A fairly general problem of this type is to find the extremum of the functional

$$J[u] = \int_R f(x_1, x_2, \ldots, x_n, u, u_{x_1}, u_{x_2}, \ldots, u_{x_n}) \, dx_1 \ldots dx_n, \tag{6.8}$$

with u given on the boundary B of a closed region R and $u \in C$, the class of functions with continuous second partial derivatives. To save a great deal of writing consider the two variable case

$$I[u] = \int_R f(x, y, u, u_x, u_y) \, dx \, dy \tag{6.9}$$

with $u \in C$ and $u(P) = g(P)$, a *given* function, for all points $P \in B$. The basic approach is the same as in section 6.1 except for replacing the integration by parts and using an extension of the basic lemma.

Let U be the extremum of I, then $U \in C$ and $U(P) = g(P)$ for $P \in B$. Consider variations around U by looking at functions $\eta(x, y) \in C$, $\eta(P) = 0$ for $P \in B$ and then

$$u = U + \varepsilon\eta \in C \quad \text{and} \quad u(P) = g(P) \quad \text{for } P \in B.$$

Substitute into (6.9) and expand by Taylor's theorem

$$I[u] = \int_R f(x, y, U, U_x, U_y) \, dx \, dy + \varepsilon \int_R \left(\eta \frac{\partial f}{\partial U} + \eta_x \frac{\partial f}{\partial U_x} + \eta_y \frac{\partial f}{\partial U_y} \right) dx \, dy + 0(\varepsilon^2).$$

Again for known U, η this is just a function $F(\varepsilon)$ of the single variable ε and since U is an extremum $F'(0) = 0$, or

$$0 = \int_R \left(\eta \frac{\partial f}{\partial U} + \eta_x \frac{\partial f}{\partial U_x} + \eta_y \frac{\partial f}{\partial U_y} \right) dx \, dy. \tag{6.10}$$

To perform the equivalent of integrating by parts, the standard integral theorem

$$\oint_B P\,dx + Q\,dy = \int_R \left(\frac{\partial Q}{\partial x} - \frac{\partial P}{\partial y}\right) dx\,dy$$

is used. First rewriting (6.10)

$$0 = \int_R \eta \left[\frac{\partial f}{\partial U} - \frac{\partial}{\partial x}\left(\frac{\partial f}{\partial U_x}\right) - \frac{\partial}{\partial y}\left(\frac{\partial f}{\partial U_y}\right)\right] dx\,dy$$
$$+ \int_R \left[\frac{\partial}{\partial x}\left(\eta \frac{\partial f}{\partial U_x}\right) + \frac{\partial}{\partial y}\left(\eta \frac{\partial f}{\partial U_y}\right)\right] dx\,dy$$

the second integral can be put to zero since the integral theorem states that

$$\int_R \left[\frac{\partial}{\partial x}\left(\eta \frac{\partial f}{\partial U_x}\right) + \frac{\partial}{\partial y}\left(\eta \frac{\partial f}{\partial U_y}\right)\right] dx\,dy = \oint_B \left(-\eta \frac{\partial f}{\partial U_y}\,dx + \eta \frac{\partial f}{\partial U_x}\,dy\right) = 0$$

and $\eta = 0$ on B. Hence

$$0 = \int_R \eta \left[\frac{\partial f}{\partial U} - \frac{\partial}{\partial x}\left(\frac{\partial f}{\partial U_x}\right) - \frac{\partial}{\partial y}\left(\frac{\partial f}{\partial U_y}\right)\right] dx\,dy.$$

This result holds for any $\eta \in C$ with $\eta = 0$ on B and the two variable extension of the basic lemma gives

$$\frac{\partial f}{\partial U} - \frac{\partial}{\partial x}\left(\frac{\partial f}{\partial U_x}\right) - \frac{\partial}{\partial y}\left(\frac{\partial f}{\partial U_y}\right) = 0.$$

The general version of this result is that *a necessary condition for an extremum of* (6.8) *is*

$$\sum_{i=1}^{n} \frac{\partial}{\partial x_n}\left(\frac{\partial f}{\partial u_{x_n}}\right) - \frac{\partial f}{\partial u} = 0. \tag{6.11}$$

A simple application of this result can be obtained by looking at the following example.

Example 39 Find the Euler equations of

$$(1) \quad \int_R (\nabla u)^2\,dx\,dy \qquad (2) \quad \int_V (u_x^2 - u_t^2)\,dx\,dt.$$

(1) $f = u_x^2 + u_y^2$ and

$$\frac{\partial f}{\partial u_x} = 2u_x, \qquad \frac{\partial f}{\partial u_y} = 2u_y$$

and (6.11) gives

$$0 = \frac{\partial}{\partial x}(2u_x) + \frac{\partial}{\partial y}(2u_y)$$

or

$$\frac{\partial^2 u}{\partial x^2} + \frac{\partial^2 u}{\partial y^2} = 0$$

which is just the Laplace equation.

(2) $f = u_x{}^2 - u_t{}^2$ and (6.11) gives

$$0 = \frac{\partial}{\partial x}(2u_x) + \frac{\partial}{\partial t}(-2u_t)$$

or

$$\frac{\partial^2 u}{\partial x^2} = \frac{\partial^2 u}{\partial t^2}$$

and this is the one dimensional wave equation.

6.4* Constraints

6.4.1* ISOPERIMETRIC PROBLEM

A key role is played by constraints in many modern problems as indicated in chapters 1 and 3. It is therefore essential to study such problems in a variational context.

Typical examples of the sort of problems involved are: (a) find the closed plane curve of given perimeter which encloses maximum area (this problem was mentioned in chapter 1) (b) minimize the total energy of a heavy chain of given length, hanging under gravity from two given points, in order to obtain its equilibrium shape. In both these problems the constraint is one of given length and hence the name *isoperimetric*. Both these problems are classical ones and will be studied in detail in chapter 7. Consider a more straightforward problem.

Example 40 Find the function u which minimizes the functional

$$J[u] = \int_0^a \left(xu'^2 + \frac{u^2}{x}\right) dx$$

with

$$u(0) = u(a) = 0 \quad \text{and} \quad 1 = \int_0^a xu^2\, dx.$$

This example is a special case (that will be solved later) of the more general problem of finding the extremum of the functional

$$I[u] = \int_a^b f(x, u, u')\, dx, \tag{6.12}$$

with $u(a) = A$, $u(b) = B$, A, B given, $u \in C^2$, and also subject to the constraint

$$L = \int_a^b g(x, u, u')\, dx. \tag{6.13}$$

To find the necessary conditions for an extremum needs a little more care than the previous cases since just putting $u = U + \varepsilon\eta$ is not quite enough. To see this, substitute into (6.13) and all that results is an equation for ε with normally a finite number of solutions. Since all variations around U are required, this initial step in the argument must first be generalized.

Let U be the extremum of (6.12) subject to (6.13) and satisfying all the other conditions. Generate the variations by considering *two* functions η, ξ satisfying η, $\xi \in C^2$, $\eta(a) = \eta(b) = \xi(a) = \xi(b) = 0$ and then

$$u = U + \varepsilon\eta + \delta\xi$$

certainly satisfies $u \in C^2$, $u(a) = A$, $u(b) = B$. Substitute into (6.13) with known U, η, ξ and this gives a relation between ε, δ, say $G(\varepsilon, \delta) = 0$. In (6.12) the substitution leads to $I[u] = F(\varepsilon, \delta)$ and the basic problem is now to find the extremum of $F(\varepsilon, \delta)$ subject to $G(\varepsilon, \delta) = 0$. This is just a Lagrange multiplier problem; Result III asks for the unconstrained extremum of $F^* = F + \lambda G$. Expanding by Taylor's theorem in the usual way

$$\begin{aligned} F^* = \int_a^b [f(x, U, U') + \lambda g(x, U, U')]\, dx \\ + \varepsilon \int_a^b [(f_U + \lambda g_U)\eta + (f_{U'} + \lambda g_{U'})\eta']\, dx \\ + \delta \int_a^b [(f_U + \lambda g_U)\theta + (f_{U'} + \lambda g_{U'})\theta']\, dx + 0(\varepsilon^2 \quad \text{or} \quad \delta^2). \end{aligned}$$

Since $\varepsilon = \delta = 0$ corresponds to the extremum then $\partial F^*/\partial\varepsilon = \partial F^*/\partial\delta = 0$ gives the required necessary conditions. Performing the integration by parts these give

$$\int_a^b \left((f_U + \lambda g_U) - \frac{d}{dx}(f_{U'} + \lambda g_{U'})\right) \eta\, dx = 0$$

$$\int_a^b \left((f_U + \lambda g_U) - \frac{d}{dx}(f_{U'} + \lambda g_{U'})\right) \theta\, dx = 0.$$

Since these equations are satisfied for independent η and θ, the constant λ is independent of the choice of trial function. Now using the normal argument gives

$$f_U - \frac{d}{dx}(f_{U'}) + \lambda\left(g_U - \frac{d}{dx}(g_{U'})\right) = 0$$

which is the Euler equation of

$$I^*[u] = \int_a^b (f + \lambda g)\,dx. \tag{6.14}$$

The above analysis establishes the result that *the extremum of* (6.12) *subject to* (6.13) *is obtained from the unconstrained extremum of* (6.14). This unconstrained extremum is obtained, of course, from the Euler equation of (6.14) with λ constant.

Example 40 The problem stated on p. 96.

Construct, with Lagrange multiplier $(-\lambda)$ for convenience,

$$J^*[u] = \int_0^a \left(xu'^2 + \frac{u^2}{x} - \lambda xu^2\right) dx.$$

The Euler equation of this functional is

$$0 = \frac{d}{dx}(2xu') - \left(\frac{2u}{x} - 2\lambda xu\right)$$

or

$$x^2u'' + xu' + (\lambda x^2 - 1)u = 0,$$

which is just the Bessel equation of order 1. Thus this constrained problem gives a method of looking at the solution of the quite complicated Bessel equation. A Rayleigh–Ritz technique used on this problem will then generate approximate solutions of the equation (see problem 8).

The general method has been applied only to functionals containing a single function of a single variable. It can, of course, be extended considerably to functionals involving multiple integrals, several functions and many constraints. Consider an example incorporating these extensions and also involving an eigenvalue problem which will be studied in more detail in chapter 8.

Example 41 (a) Find the Euler equation for the function u_1 which minimizes the functional

$$I[u] = \int_R (\nabla u)^2\,dx\,dy \tag{6.15}$$

subject to

$$\int_R u^2 \, dx \, dy = 1, \tag{6.16}$$

where $u \in C$ the class of functions with continuous second partial derivatives and $u = 0$ on the boundary B of the closed region R. (b) Find further the function u_2 which minimizes (6.15) subject to (6.16) *and* to

$$\int_R u_2 u_1 \, dx \, dy = 0, \tag{6.17}$$

$u_2 \in C$ and $u_2 = 0$ on B.

(a) Construct the functional

$$I^* = \int_R (u_x{}^2 + u_y{}^2 - \lambda_1 u^2) \, dx \, dy$$

with Euler equation

$$\nabla^2 u_1 + \lambda_1 u_1 = 0 \qquad \text{(Helmholtz equation).}$$

Now

$$\begin{aligned} I[u_1] &= \int_R \nabla u_1 \,.\, \nabla u_1 \, dx \, dy \\ &= \int_B u_1 \frac{\partial u_1}{\partial n} \, ds - \int_R u_1 \, \nabla^2 u_1 \, dx \, dy \\ &= \int_R u_1(\lambda_1 u_1) \, dx \, dy = \lambda_1 \end{aligned}$$

since $u_1 = 0$ on B, $\nabla^2 u_1 = -\lambda_1 u_1$ and (6.16) holds. Note the important result $I[u_1] = \lambda_1$.

(b) Construct the modified functional

$$I^{**} = \int_R (u_x{}^2 + u_y{}^2 - \lambda_2 u^2 - 2\mu u u_1) \, dx \, dy$$

with Euler equation

$$\nabla^2 u_2 + \lambda_2 u_2 + \mu u_1 = 0. \tag{6.18}$$

Now apply the same arguments to calculate λ_2 and μ,

$$\begin{aligned} I[u_2] &= \int_R \nabla u_2 \,.\, \nabla u_2 \, dx \, dy = \int_B u_2 \frac{\partial u_2}{\partial n} \, ds - \int_R u_2 \, \nabla^2 u_2 \, dx \, dy \\ &= \int_R u_2(\lambda_2 u_2 + \mu u_1) \, dx \, dy = \lambda_2, \end{aligned}$$

where (6.16) and (6.17) have been used. Note again that $I[u_2] = \lambda_2$. To calculate μ multiply (6.18) by u_1 and integrate over R

$$0 = \int_R u_1 \nabla^2 u_2 \, dx \, dy + \lambda_2 \int_R u_1 u_2 \, dx \, dy + \mu \int_R u_1{}^2 \, dx \, dy$$

$$= \int_R u_2 \nabla^2 u_1 \, dx \, dy + \int_B \left(u_1 \frac{\partial u_2}{\partial n} - u_2 \frac{\partial u_1}{\partial n} \right) ds + \mu.$$

Here Green's theorem has been used on the first term, (6.17) on the second term and (6.16) on the third term. Since $u_1 = u_2 = 0$ on B and since $\nabla^2 u_1 = -\lambda_1 u_1$ this equation becomes

$$\mu = -\int_R u_2(-\lambda_1 u_1) \, dx \, dy = 0.$$

Hence when u_1 and u_2 are computed exactly then $I[u_2] = \lambda_2, \mu = 0$ and (6.18) becomes

$$\nabla^2 u_2 + \lambda_2 u_2 = 0$$

which is the same Helmholtz equation. Since the class of functions C_1 from which u_1 was chosen is less restrictive than the class C_2 from which u_2 was chosen (extra equation (6.17) is required to be satisfied) then $C_1 \supset C_2$ and hence $\lambda_2 \geq \lambda_1$. In physical terms this implies that a higher eigenvalue of the Helmholtz equation has been computed.

6.4.2* NON-INTEGRAL CONSTRAINTS

Isoperimetric problems are rather restrictive since they involve only constraints which can be written as integrals. Quite simple problems can be constructed that do not fall into this category, and for control theory problems this is almost always the case.

Example 42 (a) Minimize

$$s[x, y, z] = \int_0^T (\dot{x}^2 + \dot{y}^2 + \dot{z}^2)^{\frac{1}{2}} \, dt$$

where

$$x^2 + y^2 = z^2 \tan^2 \alpha.$$

(b) Minimize

$$J[u] = \int_0^T x^2 \, dt$$

where

$$\dot{x} = f(u, x(t), t), \qquad x(0) = 0, \qquad x(T) = a$$

with a, T known.

In (a) this is just another way of writing the geodesic problem on a cone. In (b) the problem asks to minimize a quadratic cost functional in x, which itself is a solution of a differential equation (the equation of motion) dependent on the required function $u(t)$.

Both these cases are special cases of the more general problem of optimizing the functional

$$I[u, v] = \int_{t_0}^{t_1} f(t, u, v, \dot{u}, \dot{v})\, dt \tag{6.19}$$

with suitably differentiable u, v given at t_0 and t_1 and constrained by

$$g(u, v, t) = 0 \tag{6.20}$$

or

$$h(u, v, \dot{u}, \dot{v}, t) = 0. \tag{6.21}$$

These are often called respectively the holonomic and non-holonomic cases after the mechanical context in which they appear. The second problem is a little harder so concentrate on optimizing (6.19) subject to (6.20). The usual technique is quite adequate provided a little care is exercised. Let U, V be the required extreme functions and let η, ξ satisfy zero end conditions then in the usual way $u = U + \varepsilon\eta$ and $v = V + \delta\xi$ satisfy the differentiability conditions and the boundary conditions. Putting these in (6.19) and (6.20)

$$I[u, v] = \int f(t, U, V, \dot{U}, \dot{V})\, dt + \int (\varepsilon\eta f_U + \varepsilon\dot{\eta} f_{\dot{U}} + \delta\xi f_V + \delta\dot{\xi} f_{\dot{V}})\, dt$$
$$+ \text{ second order terms}$$

$$g_U \varepsilon\eta + g_V \xi\delta + (\text{second order terms}) = 0.$$

Performing the usual integration by parts, using the fact that η and ξ are zero at the end points gives

$$I[u, v] = I[U, V] + \int \left\{ \varepsilon\eta \left[f_U - \frac{d}{dt}(f_{\dot{U}}) \right] + \xi\delta \left[f_V - \frac{d}{dt}(f_{\dot{V}}) \right] \right\} dt + \cdots$$

$$= I[U, V] + \int \left\{ \left(-\frac{g_V}{g_U} \right) \left[f_U - \frac{d}{dt}(f_{\dot{U}}) \right] + f_V - \frac{d}{dt}(f_{\dot{V}}) \right\} \xi\delta\, dt + 0(\delta^2).$$

Applying the argument that for known U, V, ξ, η, $I[u, v]$ is just a function of δ, which has an extremum at $\delta = 0$, and further arguing that the same can be done for any ξ then the Euler equation takes the form

$$\left(-\frac{g_V}{g_U} \right) \left[f_U - \frac{d}{dt}(f_{\dot{U}}) \right] + \left[f_V - \frac{d}{dt}(f_{\dot{V}}) \right] = 0.$$

Putting

$$\lambda(t) = -\frac{[f_U - (d/dt)(f_{\dot{U}})]}{g_U}$$

finally gives

$$\frac{d}{dt}(f_{\dot{U}}) - f_U - \lambda(t)g_U = 0$$

$$\frac{d}{dt}(f_{\dot{V}}) - f_V - \lambda(t)g_V = 0.$$

Thus *optimizing* (6.19) *subject to* (6.20) *is equivalent to optimizing the unconstrained functional*

$$I^*[u, v] = \int_{t_0}^{t_1} [f + \lambda(t)g]\, dt. \tag{6.22}$$

It should be noted that the additional complication is that the Lagrange multiplier λ is now a function of t. For the non-holonomic case, optimizing (6.19) subject to (6.21) the similar result can be proved: *the extremum of* (6.19) *subject to* (6.21) *is obtained from the unconstrained extremum of*

$$I^{**}[u, v] = \int_{t_0}^{t_1} [f + \mu(t)h]\, dt. \tag{6.23}$$

Example 42 Parts (a) and (b) cited on p. 100.

(a) Construct

$$s^*[x, y, z] = \int [(\dot{x}^2 + \dot{y}^2 + \dot{z}^2)^{\frac{1}{2}} + \lambda(t)(x^2 + y^2 - z^2 \tan \alpha)]\, dt.$$

The Euler equations are

$$0 = \frac{d}{dt}\left(\frac{\dot{x}}{\dot{s}}\right) - 2\lambda x$$

$$0 = \frac{d}{dt}\left(\frac{\dot{y}}{\dot{s}}\right) - 2\lambda y$$

$$0 = \frac{d}{dt}\left(\frac{\dot{z}}{\dot{s}}\right) + 2\lambda z \tan^2 \alpha,$$

where the fact that $\dot{s}^2 = \dot{x}^2 + \dot{y}^2 + \dot{z}^2$ has been used. These equations can be rearranged by eliminating λ and using 'dash' to denote differentiation with respect to s,

$$0 = yx'' - xy'' = \frac{d}{ds}(yx' - xy')$$

$$0 = xz'' + zx'' \tan^2 \alpha,$$

which can be solved together with the constraint equation

$$x^2 + y^2 = z^2 \tan^2 \alpha.$$

Note that this is an equivalent formulation of example 38, but it is not in such a convenient form for the final integrations.

(b) Again a modified functional is formed

$$J^*[x, u] = \int_0^T \{x^2 + \mu(t)[\dot{x} - f(u, x, t)]\}\, dt$$

and the Euler equations are

$$0 = \frac{d}{dt}[\mu(t)] - (2x - \mu f_x)$$

$$0 = \mu f_u.$$

Since $x \not\equiv 0$ and $\mu \not\equiv 0$ then the condition for the extremum is $f_u = 0$ together with the differential equation $\dot{x} = f(u, x, t)$.

Example 43 Find the extremum of the functional

$$\int_a^b (\tfrac{1}{2}y''^2 - ky)\, dx.$$

This problem was considered by approximate methods in chapter 4 (see example 27). None of the methods developed so far has dealt with any sort of higher derivatives, but the constrained problems now studied give an opportunity to remedy this.

By putting $y' = z$, the problem is changed to finding the extremum of

$$I[y, z] = \int_a^b [\tfrac{1}{2}z'^2 - k(x)y]\, dx$$

subject to $y' - z = 0$. The modified functional is

$$I^*[y, z] = \int_a^b [(\tfrac{1}{2}z'^2 - ky) + \lambda(x)(y' - z)]\, dx$$

with Euler equations

$$0 = z'' + \lambda$$

$$0 = \lambda' + k(x)$$

together with $y' = z$. Combining these three equations by eliminating λ and z produces the Euler equation

$$y'''' = k(x).$$

Compare the solution of this equation with example 27.

Example 44 An aircraft in level flight has natural speed V and is subject to an additional constant wind of speed U ($<V$). Find the closed flight path of the aircraft which maximizes the area enclosed in a given time.

This problem can be interpreted as the flight path to maximize the area inspected in an aerial reconnaissance. With axes as shown in figure 6.2 the equations of motion are

$$\dot{x} = U + V\cos\theta, \qquad \dot{y} = V\sin\theta,$$

while the area can be written

$$A = \int_R dx\,dy = \tfrac{1}{2}\int_B (-y\,dx + x\,dy) = \tfrac{1}{2}\int_0^T (-y\dot{x} + x\dot{y})\,dt,$$

where T is the given time. The middle equality in this expression is obtained from the integral theorem

$$\int_R \left(\frac{\partial Q}{\partial x} - \frac{\partial P}{\partial y}\right) dx\,dy = \oint_B P\,dx + Q\,dy$$

with $P = -y$ and $Q = x$.

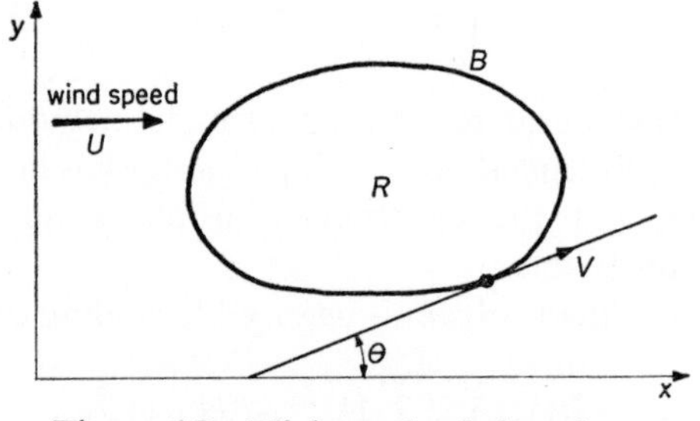

Figure 6.2 Flight path of aircraft.

Thus the basic problem is to maximize A subject to the equations of motion. Constructing the modified functional

$$A^* = \int [(-y\dot{x} + x\dot{y}) + \lambda(t)(\dot{x} - U - V\cos\theta) + \mu(t)(\dot{y} - V\sin\theta)]\,dt$$

the Euler equations are deduced as

$$0 = \frac{d}{dt}(-y + \lambda) - \dot{y}$$

$$0 = \frac{d}{dt}(x + \mu) + \dot{x}$$

$$0 = -(\lambda V\sin\theta - \mu V\cos\theta).$$

The first two of these equations give $2y = \lambda + \alpha$ and $2x = -\mu + \beta$, where α and β are arbitrary constants. The third equation gives

$$\tan\theta = \frac{\mu}{\lambda} = \frac{\beta - 2x}{2y - \alpha}.$$

Defining

$$r^2 = (2y - \alpha)^2 + (\beta - 2x)^2$$

$$\begin{aligned} 2r\dot{r} &= 4(2y - \alpha)\dot{y} - 4(\beta - 2x)\dot{x} \\ &= [4(2y - \alpha)V\sin\theta - 4(\beta - 2x)V\cos\theta] - 4(\beta - 2x)U \end{aligned}$$

or

$$\frac{dr}{dt} = -\frac{(\beta - 2x)2U}{r} = -2U\sin\theta = -\frac{2U}{V}\frac{dy}{dt}.$$

Integrating this final equation gives $r = -(2U/V)y + C$ which is just the ellipse

$$(2y - \alpha)^2 + (\beta - 2x)^2 = \left[C - \left(\frac{2U}{V}\right)y\right]^2.$$

6.5* Natural Boundary Conditions

In all the problems considered so far the boundary conditions have been specified carefully as u given on the boundary. This is of course too restrictive and it must be possible to treat other boundary conditions or even to leave these conditions unspecified. As a start consider the standard functional with a single function of one variable with less restrictive boundary conditions: find the extremum of

$$J[u] = \int_a^b f(x, u, u')\,dx$$

with $u \in C^2$ and

$$u(a) = A \qquad \text{but } u(b) \text{ unspecified.}$$

Since the right-hand end is unspecified the class of admissible functions is much wider than the standard problem considered in section 6.1.

The method of analysing this problem is similar to the usual method. Let $U \in C^2$ be the extremum and hence $U(a) = A$; let $\eta \in C^2$ and $\eta(a) = 0$ but with no restriction on $\eta(b)$ for the moment. The function

$$u = (U + \varepsilon\eta) \in C^2 \quad \text{and} \quad u(a) = A$$

as required. Substituting into the functional gives in the usual way $J[U + \varepsilon\eta] = F(\varepsilon)$ and with many of the steps now omitted

$$F'(\varepsilon) = \int_a^b \left[\eta \frac{\partial f}{\partial U} + \eta' \left(\frac{\partial f}{\partial U'} \right) \right] dx + 0(\varepsilon).$$

Since $\varepsilon = 0$ is an extremum of the problem, $F'(0) = 0$ or

$$0 = \int_a^b \left[\eta \frac{\partial f}{\partial U} + \eta' \left(\frac{\partial f}{\partial U'} \right) \right] dx$$

$$= \int_a^b \eta \left[\frac{\partial f}{\partial U} - \frac{d}{dx} \left(\frac{\partial f}{\partial U'} \right) \right] dx + \eta(b) \frac{\partial f}{\partial U'} \bigg|_{x=b}. \tag{6.24}$$

The function η is now chosen to belong to the more restrictive class with $\eta(b) = 0$. This is all that is needed to apply the basic lemma to give the usual Euler equation

$$\frac{d}{dx} \left(\frac{\partial f}{\partial U'} \right) - \frac{\partial f}{\partial U} = 0.$$

The statement (6.24) is, however, true for all $\eta \in C^2$ with $\eta(a) = 0$ and in general $\eta(b) \neq 0$. Thus the remaining term in (6.24) must be zero

$$\frac{\partial f}{\partial U'} = 0 \quad \text{at } x = b. \tag{6.25}$$

This is called a *natural boundary condition.*

Example 45 Find the extremum of the functional

$$J[u] = \int_0^1 \left(x u'^2 + \frac{u^2}{x} \right) dx$$

with $u(0) = 0$ and $u(1)$ unspecified.

The Euler equation is

$$0 = \frac{d}{dx}(2xu') - \frac{2u}{x}$$

or

$$0 = x^2 u'' + xu' - u. \tag{6.26}$$

The natural boundary condition is

$$\frac{\partial f}{\partial u'} = 2xu' = 0 \quad \text{at } x = 1.$$

Hence the problem is equivalent to solving the differential equation (6.26) subject to the boundary conditions $u(0) = 0$, $u'(1) = 0$.

Likewise for many problems of this type derivative boundary conditions can be dealt with. An interesting problem that illustrates the use of natural boundary conditions in a multiple integral situation occurs in electrostatics.

Example 46 The electric intensity, $\mathbf{e}$, in a capacitor satisfies

$$\operatorname{div} \mathbf{e} = 0 \quad \text{in } R,$$

where R is the region between two closed non-intersecting surfaces S_1 and S_2. The surfaces are held at charges Q and $-Q$ respectively which implies, using Gauss's theorem

$$\int_{S_1} \mathbf{e} \,.\, d\mathbf{S} = 4\pi Q, \qquad \int_{S_2} \mathbf{e} \,.\, d\mathbf{S} = -4\pi Q. \tag{6.27}$$

Find the equilibrium distribution for $\mathbf{e}$ by minimizing the energy

$$\frac{1}{8\pi} \int_R \mathbf{e}^2 \, dV.$$

The $\operatorname{div} \mathbf{e} = 0$ constraint can be taken care of by using a Lagrange multiplier $\lambda(x, y, z)$, while the two integral boundary constraints can be dealt with using constant Lagrange multipliers μ and ν (this can easily be deduced). Minimize therefore

$$J^*[\mathbf{e}] = \int_R (\tfrac{1}{2}\mathbf{e}^2 + \lambda \operatorname{div} \mathbf{e}) \, dV + \mu \int_{S_1} \mathbf{e} \,.\, d\mathbf{S} + \nu \int_{S_2} \mathbf{e} \,.\, d\mathbf{S}.$$

The integrals round the boundaries complicate matters so work again from first principles. Let $\mathbf{E}$ be the required extremum and $\boldsymbol{\eta}$ a suitably differentiable vector function. Consider the variations

$$\mathbf{e} = \mathbf{E} + \varepsilon \boldsymbol{\eta}$$

$$J^*[\mathbf{e}] = F(\varepsilon) = \int_R (\tfrac{1}{2}\mathbf{E}^2 + \varepsilon \mathbf{E} \,.\, \boldsymbol{\eta} + \lambda \operatorname{div} \mathbf{E} + \varepsilon\lambda \operatorname{div} \boldsymbol{\eta}) \, dV$$
$$+ \mu \int_{S_1} (\mathbf{E} + \varepsilon\boldsymbol{\eta}) \,.\, d\mathbf{S} + \nu \int_{S_2} (\mathbf{E} + \varepsilon\boldsymbol{\eta}) \,.\, d\mathbf{S} + 0(\varepsilon^2).$$

Satisfying the requirement that $\mathbf{E}$ is an extremum or $F'(0) = 0$ leads to

$$0 = \int_R (\mathbf{E} \,.\, \boldsymbol{\eta} + \lambda \operatorname{div} \boldsymbol{\eta}) \, dV + \mu \int_{S_1} \boldsymbol{\eta} \,.\, d\mathbf{S} + \nu \int_{S_2} \boldsymbol{\eta} \,.\, d\mathbf{S}.$$

This expression can be further manipulated by the use of the vector identity

$$\lambda \operatorname{div} \boldsymbol{\eta} = \operatorname{div} (\lambda\boldsymbol{\eta}) - \nabla\lambda \,.\, \boldsymbol{\eta}$$

and by the divergence theorem to give

$$0 = \int_R (\mathbf{E} - \nabla\lambda) \cdot \boldsymbol{\eta}\, dV + \int_{S_1} (\mu - \lambda)\boldsymbol{\eta} \cdot \mathbf{dS} + \int_{S_2} (\nu - \lambda)\boldsymbol{\eta} \cdot \mathbf{dS}.$$

For the moment consider the restrictive condition $\boldsymbol{\eta} = 0$ on S_1 and S_2 then an extension of the basic lemma states that if

$$0 = \int_R (\mathbf{E} - \nabla\lambda) \cdot \boldsymbol{\eta}\, dV$$

for *all* $\boldsymbol{\eta}$ with $\boldsymbol{\eta} = 0$ on the boundary then

$$\mathbf{E} = \nabla\lambda.$$

(In this extension it is assumed that all functions are sufficiently differentiable.) Further let $\boldsymbol{\eta} = 0$ on S_1 but $\boldsymbol{\eta} \neq 0$ on S_2 then the implication is that $\lambda = \nu$ on S_2 and similarly $\lambda = \mu$ on S_1.

Thus the electrostatic problem can be reduced to solving

$$\text{div}\,\mathbf{E} = \nabla^2\lambda = 0$$

with

$$\lambda = \mu \quad \text{on } S_1 \qquad \text{and} \qquad \lambda = \nu \quad \text{on } S_2,$$

these boundary conditions being the *natural* ones; the actual values of the constants μ and ν can be calculated from the integrals (6.27).

PROBLEMS

1. Use the *first principle* method of section 6.1 to obtain the Euler equation of

$$\int_0^{\frac{1}{2}\pi} (y'^2 - y^2)\, dx, \qquad y(0) = 0, y(\tfrac{1}{2}\pi) = 1.$$

2. Find the Euler equations of (cf. problems in chapter 5)

(i) $$\int_1^2 x(y'^2 - y^2)\, dx, \qquad y(1) = 1, y(2) = 2.$$

(ii) $$\int_a^b \frac{(1 + y'^2)^{\frac{1}{2}}}{y}\, dx, \qquad y(a), y(b) \text{ given}.$$

(iii) $$\int_0^T \left(\frac{1}{2}\dot{r}^2 + \frac{1}{2}r^2\dot{\theta}^2 + \frac{\mu}{r}\right) dt, \qquad r, \dot{r}, \theta, \dot{\theta} \text{ given at } t = 0.$$

3. The cost of laying cable of length ds at a height z is $z\,ds$. The contours of z on a map in the (x, y) plane take the form $z = 1/(1 + x^2 + y^2) = \text{constant}$.

Show that the total cost of laying the cable from $(-a, 0)$ to $(a, 0)$ is

$$C[r] = \int_{-\pi}^{0} \frac{[(dr/d\theta)^2 + r^2]^{\frac{1}{2}}}{1 + r^2} \, d\theta,$$

where (r, θ) are the usual polar coordinates in the plane. Find the minimum cost and the path taken. (Do not attempt the final integration.)

4. The flight path of an aircraft is required between two places distance a apart and both at sea level. The cost of flying the aircraft a distance ds at a height h above sea level is $\exp(-h/H)\, ds$. Find the flight path giving overall minimum cost.

5. Using the parametric representation of a sphere as

$$x = a \sin\theta \cos\varphi, \qquad y = a \sin\theta \sin\varphi, \qquad z = a\cos\theta$$

construct a formula for the length of the curve $\varphi = \varphi(\theta)$ between two suitable points on the sphere. Find the Euler equation and solve it.

6. Find the Euler equations of the functionals

(i) $\displaystyle\iint_R (1 + u_x{}^2 + u_y{}^2)^{\frac{1}{2}} \, dx\, dy$ (minimum surface area problem)

(ii) $\displaystyle\int_V [(\nabla u)^2 - g(x, y, z)u] \, dV.$

7. Minimize the functional

$$J[u] = \int_0^1 u'^2 \, dx \quad \text{subject to} \quad \int_0^1 u^2 \, dx = 1$$

with $u(0) = u(1) = 0$. If U is the extremum, evaluate $J[U]$.

8. In example 40 show that $J[U] = \lambda$, where λ is the Lagrange multiplier, and use the Rayleigh–Ritz procedure with $u = Ax(a - x)$ to estimate λ.

9. Find the minimum of the functional

$$\int_{t_1}^{t_2} (\dot{x}^2 + \dot{y}^2 + \dot{z}^2)^{\frac{1}{2}} \, dt$$

subject to $x^2 + y^2 + z^2 = a^2$. (cf. problem 5)

10. Find the function $u(t)$ which gives minimum cost

$$C[u] = \int_0^1 (u^2 + y^2) \, dt$$

where $\dot{y} = u(t)$ and $y(0) = 1$. Note that a natural boundary condition is required on the Lagrange multiplier at $t = 1$.

11. By putting $u' = z$ find the Euler equation of

$$\int f(x, u, u', u'') \, dx.$$

12. Show that optimizing the functional

$$J[u] = \int_0^1 (u'^2 - u^2)\, dx - u^2(1)$$

subject to $u(0) = 1$, $u(1)$ unspecified is equivalent to solving

$$u'' + u = 0$$

with the boundary conditions $u(0) = 1$ and $u'(1) = u(1)$. Test this with a suitable trigonometric or polynomial approximation used together with the Rayleigh–Ritz procedure.

13. A circular membrane (a drum) satisfies the wave equation

$$\nabla^2 u = \frac{1}{c^2}\frac{\partial^2 u}{\partial t^2},$$

where u is the displacement and c is a constant. The edge of the drum is clamped, $u = 0$, over part of the boundary and left free over the other part. Find the natural boundary condition on the free part by considering the functional

$$\iiint [c^2(\nabla u)^2 - u_t^2]\, dx\, dy\, dt.$$

Chapter VII

Classical Applications

7.1 Brachistochrone

Out of several problems that could be called classical one or two have been selected for special mention. Perhaps the best known of these is the brachistochrone problem which was first introduced in example 34 in section 5.3. There an approximate technique was tried and it was found to produce very difficult equations to solve. Fortunately this problem can be solved exactly using the Euler equation established in the last chapter. The problem is restated in the following example.

Example 47 Axes Oxy are set up with Ox horizontal and Oy vertically downwards. A smooth wire joins the points $(0, 0)$ and $A(a, b)$. A bead, initially at rest, slides down the wire from O to A under gravity. It is required to find the shape of the wire that minimizes the time of descent.

The energy equation

$$\tfrac{1}{2}mv^2 = mgy$$

can be integrated as in example 34 to give the time of descent T as

$$T = \frac{1}{(2g)^{\frac{1}{2}}} \int_0^a \left(\frac{1 + y'^2}{y}\right)^{\frac{1}{2}} dx,$$

$$y(0) = 0, \qquad y(a) = b.$$

The minimum of the functional $T[y]$ is obtained by writing the Euler equation in its *integrated form* (6.1) since the functional is independent explicitly of x,

$$C = y' \frac{\partial f}{\partial y'} - y = y' \frac{y'}{[y(1 + y'^2)]^{\frac{1}{2}}} - \left(\frac{1 + y'^2}{y}\right)^{\frac{1}{2}}.$$

Rearranging this equation gives

$$(1 + y'^2)y = K,$$

where K is a new arbitrary constant ($K = 1/C^2$). This differential equation can be integrated once more using the substitution $y = K \sin^2 \theta$ to give

$$x - A = K(\theta - \tfrac{1}{2} \sin 2\theta).$$

The second arbitrary constant $A = 0$ since when $x = y = 0$, $\theta = 0$ also. The constant K is evaluated from $y(a) = b$. Putting $B = \frac{1}{2}K$ and $\varphi = 2\theta$ brings the equations for x and y into the standard form

$$x = B(\varphi - \sin \varphi), \qquad y = B(1 - \cos \varphi),$$

which is the usual parametric equation of a *cycloid.*

This result is a very old one and was first produced by Bernouilli at the end of the seventeenth century. It is one of the few physical problems of this type for which an exact solution can be produced and it is this exact solution and the historical interest which makes the problem worthy of study.

7.2* Minimum Energy

In this section it is the aim to establish that a conservative system is in stable equilibrium at the minimum of the potential energy. This result has been used several times, in examples 16, 27, 46 for instance.

First look at a dynamical system defined uniquely by n generalized independent coordinates $q_1, q_2, \ldots, q_n$, then for each particle, of mass m_i at $\mathbf{r}_i$, of the system

$$\mathbf{r}_i = \mathbf{r}_i(q_1, q_2, \ldots, q_n, t).$$

Since

$$\dot{\mathbf{r}}_i = \sum_{j=1}^{n} \frac{\partial \mathbf{r}_i}{\partial q_j} \dot{q}_j + \frac{\partial \mathbf{r}_i}{\partial t}$$

then differentiating this with respect to $\dot{q}_j$ gives

$$\frac{\partial \dot{\mathbf{r}}_i}{\partial \dot{q}_j} = \frac{\partial \mathbf{r}_i}{\partial q_j}.$$

This is a result that will be required in the later manipulation. The equations of motion of the system take the form

$$m_i \ddot{\mathbf{r}}_i = \mathbf{F}_i + \mathbf{F}'_i, \tag{7.1}$$

where $\mathbf{F}_i$ is the external force and $\mathbf{F}'_i$ the internal force on a typical particle.

It is assumed that the internal forces do not affect the external motion and hence when summed for the whole system they give zero contribution. The external forces are assumed to be derived from a potential $V(q_1, q_2, \ldots, q_n)$, since the system is conservative, thus $\mathbf{F}_i = -\nabla_i V$ where ∇_i denotes the gradient taken with respect to $\mathbf{r}_i = (x_i, y_i, z_i)$. Dotting (7.1) with $\partial \mathbf{r}_i/\partial q_j$ and summing over all particles

$$\sum m_i \frac{\partial \mathbf{r}_i}{\partial q_j} \cdot \ddot{\mathbf{r}}_i = \sum \frac{\partial \mathbf{r}_i}{\partial q_j} \cdot \mathbf{F}_i + \sum \frac{\partial \mathbf{r}_i}{\partial q_j} \cdot \mathbf{F}'_i$$

or

$$\sum m_i \frac{d}{dt}\left(\frac{\partial \mathbf{r}_i}{\partial q_j} \cdot \dot{\mathbf{r}}_i\right) - \sum m_i \frac{d}{dt}\left(\frac{\partial \mathbf{r}_i}{\partial q_j}\right) \cdot \dot{\mathbf{r}}_i = -\sum \frac{\partial \mathbf{r}_i}{\partial q_j} \cdot \nabla_i V$$

$$\frac{d}{dt}\left(\sum m_i \frac{\partial \dot{\mathbf{r}}_i}{\partial \dot{q}_j} \cdot \dot{\mathbf{r}}_i\right) - \sum m_i \frac{\partial \dot{\mathbf{r}}_i}{\partial q_j} \cdot \dot{\mathbf{r}}_i = -\frac{\partial V}{\partial q_j}$$

$$\frac{d}{dt}\left[\frac{\partial}{\partial \dot{q}_j}\left(\sum \tfrac{1}{2} m_i \dot{\mathbf{r}}_i^{\,2}\right)\right] - \frac{\partial}{\partial q_j}\left(\sum \tfrac{1}{2} m_i \dot{\mathbf{r}}_i^{\,2}\right) = -\frac{\partial V}{\partial q_j}.$$

Hence

$$\frac{d}{dt}\left(\frac{\partial T}{\partial \dot{q}_j}\right) - \frac{\partial T}{\partial q_j} = -\frac{\partial V}{\partial q_j}, \qquad (j = 1, 2, \ldots, n), \tag{7.2}$$

where T is the usual kinetic energy. These equations are called the *Lagrange equations* for a conservative holonomic system. For a proof of the corresponding results for other systems, the reader is referred to any advanced text on classical mechanics.

The minimum energy theorem is now straightforward to establish. In equilibrium the values of the generalized coordinates $q_1, \ldots, q_n$ take constant values therefore the velocities $\dot{\mathbf{r}}_i$ of the particles of the system are zero and hence $T = \Sigma \frac{1}{2} m_i \dot{\mathbf{r}}_i^{\,2} = 0$. Thus from (7.2) if $T = 0$

$$\frac{\partial V}{\partial q_j} = 0 \qquad (j = 1, 2, \ldots, n) \tag{7.3}$$

and hence the equilibrium values of q_j are given by the solution of the n equations (7.3). These of course are just the necessary conditions for an extremum of the potential energy $V(q_1, q_2, \ldots, q_n)$. It is physically reasonable and indeed it can be proved that for stable equilibrium this extremum is a minimum otherwise it is in unstable equilibrium (see problem 3).

This result can be generalized to situations where the number of coordinates becomes infinite or even continuous. This latter case is the one that has already been used and just the case required for variational problems. One interesting classical problem that can be solved in this way is the following.

Example 48 A heavy uniform chain of length L hangs under gravity from two points at the same level. Find the equilibrium shape of the chain.

Setting up axes as shown in figure 7.1 the potential energy of the element, ds, of the chain is $-y(\rho g\, ds)$, where ρ is the density of the chain. The total energy is therefore

$$E = -\int \rho g y \, ds$$

and the total length L is

$$L = \int ds.$$

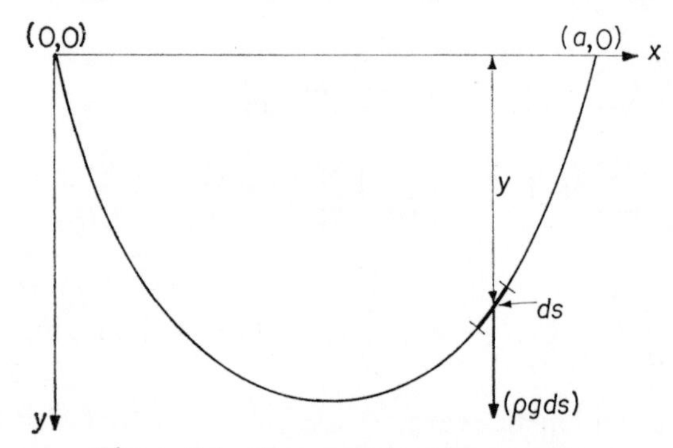

Figure 7.1 Heavy chain under gravity.

Recalling that $ds^2 = dx^2 + dy^2$ these can be rewritten

$$E[y] = -\rho g \int_0^a y(1 + y'^2)^{\frac{1}{2}} \, dx \tag{7.4}$$

subject to

$$L = \int_0^a (1 + y'^2)^{\frac{1}{2}} \, dx \tag{7.5}$$

and to pass through the end points

$$y(0) = y(a) = 0. \tag{7.6}$$

To obtain the equilibrium shape of the chain it is necessary to minimize (7.4) with respect to the function y subject to (7.5) and (7.6). This problem is just the isoperimetric problem of section 6.4.1 and can be solved using a Lagrange multiplier, λ, on (7.5). Thus the unconstrained extremum of

$$E^*[y] = -\rho g \int_0^a (y + \lambda)(1 + y'^2)^{\frac{1}{2}} \, dx$$

is required. Since the integrand is explicitly independent of x the first integral (6.1) of the Euler equation can be used,

$$\text{constant} = y' \frac{\partial f}{\partial y'} - f = -\frac{(y + \lambda)}{(1 + y'^2)^{\frac{1}{2}}}.$$

This can be converted to the differential equation

$$\frac{dy}{dx} = [K^2(y + \lambda)^2 - 1]^{\frac{1}{2}},$$

which can be solved using the substitution $K(y + \lambda) = \cosh \theta$, to give

$$\cosh K(x + \alpha) = K(y + \lambda).$$

This is the usual solution of the problem, the curve being called a *catenary*. The three unknown constants K, α, λ are obtained from the two conditions in (7.6) and the constraint in (7.5).

Although the minimum energy was proved only for a very restricted case the idea is well used in many physical situations. For instance in thermodynamics one of the basic methods of calculation is to minimize some potential, the Gibbs or Helmholtz free energies for instance, to obtain the stable equilibrium situation. Similarly in electrostatics and elastostatics corresponding results hold, once the total energy can be calculated the equilibrium configuration can then be computed by minimization (see example 46). Even in fluid mechanics where non-equilibrium situations are studied there is an important minimum energy theorem due to Kelvin (see problem 6). It can be seen that such theorems in a variety of physical situations give the possibility of solving for the equilibrium configuration by a minimization technique. This is often of considerable value since the approximate methods of chapter 5 or the hill climbing methods of chapter 2 give another method of attack, which in particular circumstances, may be a very convenient one.

7.3* Hamilton's Principle

Very closely related to the Lagrange equations (7.2) is a variational formulation due to Hamilton. Again consider only the simplest system of a conservative holonomic one; that is the system is defined uniquely by n independent generalized coordinates $q_1, q_2, \ldots, q_n$, a potential energy $V(q_1, \ldots, q_n)$ exists and the kinetic energy is given by $T = T(q_1, \ldots, q_n, \dot{q}_1, \ldots, \dot{q}_n, t)$. Consider the Lagrangian $L = T - V$ then the functional between the times t_0 and t_1

$$J[q_1, \ldots, q_n] = \int_{t_0}^{t_1} L\, dt \tag{7.7}$$

has an extremum given by the Euler equations

$$\frac{d}{dt}\left(\frac{\partial L}{\partial \dot{q}_i}\right) - \frac{\partial L}{\partial q_i} = 0 \qquad i = 1, \ldots, n$$

or

$$\frac{d}{dt}\left(\frac{\partial T}{\partial \dot{q}_i}\right) - \frac{\partial T}{\partial q_i} = -\frac{\partial V}{\partial q_i} \qquad i = 1, \ldots, n \tag{7.8}$$

which is identical with the Lagrange equations (7.2). Thus if $q_1, \ldots, q_n$ are defined at t_0 and t_1 then out of all possible functions $q_1(t), \ldots, q_n(t)$ which are sufficiently differentiable and satisfy the end conditions, the *actual motion* proceeds according to the functions or paths which minimize (7.7). This is *Hamilton's principle.*

Example 49 Find the path of a projectile which passes through (0, 0) at zero time and $(a, 0)$ at time τ, where axes Oxy are chosen with Ox horizontal and Oy vertically upwards.

Now

$$T = \tfrac{1}{2}m(\dot{x}^2 + \dot{y}^2), \qquad V = mgy$$

and hence it is required to minimize

$$J[x, y] = \int_0^\tau [\tfrac{1}{2}m(\dot{x}^2 + \dot{y}^2) - mgy]\, dt. \tag{7.9}$$

The Euler equations of this functional (using the standard notation) are

$$0 = \frac{d}{dt}\left(\frac{\partial f}{\partial \dot{x}}\right) - \frac{\partial f}{\partial x} = \frac{d}{dt}(m\dot{x})$$

$$0 = \frac{d}{dt}\left(\frac{\partial f}{\partial \dot{y}}\right) - \frac{\partial f}{\partial y} = \frac{d}{dt}(m\dot{y}) + mg$$

and the Euler–Lagrange equations become the usual equations of motion

$$m\ddot{x} = 0, \qquad m\ddot{y} = -mg.$$

These equations can be solved easily but an alternative method is to use simple approximate solutions in (7.9). For instance try

$$x = as, \qquad y = A \sin \pi s,$$

where $s = t/\tau$. The boundary conditions are clearly satisfied, so substituting into (7.9) gives

$$J[x, y] = \tfrac{1}{2}\left(\frac{m}{\tau}\right)(a^2 + \tfrac{1}{2}A^2\pi^2) - \frac{2mg\tau A}{\pi}.$$

Evaluating the extremum by putting $\partial J/\partial A = 0$ provides the solution $A = 4\tau^2 g/\pi^3$. For this simple problem it is clearly possible to construct the exact solution but for more complicated problems an approximate solution of the above type can be very useful, particularly when a great deal of physical or geometrical intuition is fed into the trial functions. Such solutions can be quite accurate; in the present case, for example, the exact maximum of y is $0.125g\tau^2$ while the approximate value is $0.129g\tau^2$.

The statement of Hamilton's principle can be thought of as replacing Newton's second law of motion. In certain circumstances it can be a convenient alternative and it will be shown in use in later chapters. The principle can be cast into another form which will give the equations of motion in Hamiltonian form rather than Lagrangian form (see problem 9). This formulation is useful in control theory and further discussion will be left until chapter 11.

7.4* Geodesics

Numerous attacks have already been made on geodesics in previous chapters: examples 6 and 42, problems 5 and 9 of chapter 6. It was found that the problem could be approached in several ways. The major question left was whether the problem can be solved generally; at least, can the equations for solution be written down? Such a general approach to geodesics is possible but is quite difficult since it requires considerable skill with suffices and with some of the matrix algebra. The starting point is to assume that the element of arc length, ds, in the space being considered can be written

$$ds^2 = \sum_{i,j} g_{ij}(x_1, x_2, \ldots, x_n)\, dx_i\, dx_j. \tag{7.10}$$

(Note that for geodesics in a plane (4.1) and on a cone (6.7) ds^2 took the form $ds^2 = dx^2 + dy^2$, $ds^2 = \sec^2 \alpha\, dz^2 + \tan^2 \alpha z^2\, d\theta^2$ respectively.) The matrix with elements g_{ij} is assumed to be symmetric (without any loss of generality) and positive definite (see section 2.4.2, 2.4.3) since arc length is an essentially positive quantity. If t is a parameter along a curve in the space under consideration then the arc length along the curve from $t = t_0$ to $t = t_1$ can be computed from

$$s[x_1, x_2, \ldots, x_n] = \int_{t_0}^{t_1} \left(\sum g_{ij}\dot{x}_i\dot{x}_j\right)^{\frac{1}{2}} dt. \tag{7.11}$$

The Euler equations, giving necessary conditions for an extremum, can now be constructed as

$$0 = \frac{d}{dt}\left(\frac{\partial f}{\partial \dot{x}_k}\right) - \frac{\partial f}{\partial x_k} = \frac{d}{dt}\left\{\frac{\sum_j g_{kj}\dot{x}_j}{\left(\sum g_{ij}\dot{x}_i\dot{x}_j\right)^{\frac{1}{2}}}\right\} - \frac{\sum_{ij} (\partial g_{ij}/\partial x_k)\dot{x}_i\dot{x}_j}{\left(\sum g_{ij}\dot{x}_i\dot{x}_j\right)^{\frac{1}{2}}}.$$

Using the arc length itself as the parameter along the curve gives from (7.10)

$$\frac{d}{ds} \equiv \frac{1}{(\Sigma\, g_{ij}\dot{x}_i\dot{x}_j)^{\frac{1}{2}}}\frac{d}{dt}$$

and some simplification results

$$0 = \frac{d}{ds}\left(\sum_j g_{kj}x'_j\right) - \tfrac{1}{2}\sum_{i,j}\frac{\partial g_{ij}}{\partial x_k}x'_i x'_j,$$

where dash denotes differentiation with respect to s. Differentiating out the first term gives

$$0 = \sum_j g_{kj}x''_j + \sum_{i,j}\left(\frac{\partial g_{kj}}{\partial x_i} - \frac{1}{2}\frac{\partial g_{ij}}{\partial x_k}\right)x'_i x'_j.$$

Defining the *Christoffel symbol of the first kind* as

$$[k, ij] = \tfrac{1}{2}\left(\frac{\partial g_{kj}}{\partial x_i} + \frac{\partial g_{ki}}{\partial x_j} - \frac{\partial g_{ij}}{\partial x_k}\right)$$

the equation can be rewritten in the more compact form

$$0 = \sum_j g_{kj}x''_j + \sum_{i,j}[k, ij]x'_i x'_j. \tag{7.12}$$

These equations are still not in the most convenient form and can be simplified by recalling that the matrix (g_{kj}) was positive definite and hence has an inverse (g^{pk}) such that

$$\sum_k g^{pk}g_{kj} = \delta_{pj}, \quad \text{all } p, j.$$

Premultiplying the (7.12) by the g^{pk} and summing over k, removes the first summation

$$0 = x''_p + \sum_{i,j}\left(\sum_k g^{pk}[k, ij]\right)x'_i x'_j.$$

The bracketed term is normally called the *Christoffel symbol of the second kind*

$$\begin{Bmatrix} p \\ i \quad j \end{Bmatrix} = \sum_k g^{pk}[k, ij]$$

and the geodesic equations take the final form

$$0 = x''_p + \sum_{i,j} \begin{Bmatrix} p \\ i \quad j \end{Bmatrix} x'_i x'_j, \qquad p = 1, \ldots, n. \tag{7.13}$$

To obtain these equations for a particular g_{ij} the basic problem is to evaluate the Christoffel symbols (note that it is symmetric in i and j) which only depend on the g_{ij}. The remaining problem is then to solve the differential equations obtained; this usually turns out to be a particularly difficult task.

Example 50 Evaluate the Christoffel symbols for the metric given by the distance on the surface of a paraboloid of revolution $x^2 + y^2 = 2z$ and hence deduce the geodesic equations.

Using cylindrical polar coordinates $x = r\cos\theta$, $y = r\sin\theta$, $z = z$ the equation becomes $r^2 = 2z$ and

$$\begin{aligned} ds^2 &= dr^2 + r^2\,d\theta^2 + dz^2 \\ &= (1 + r^2)\,dr^2 + r^2\,d\theta^2. \end{aligned}$$

Taking $r = x_1$, $\theta = x_2$ in the above notation the matrix g_{ij} is diagonal with components

$$g_{11} = 1 + r^2, \qquad g_{22} = r^2, \qquad g_{21} = g_{12} = 0$$

and

$$g^{11} = \frac{1}{1 + r^2}, \qquad g^{22} = \frac{1}{r^2}, \qquad g^{21} = g^{12} = 0.$$

The only non-zero Christoffel symbols of the first kind are

$$[1, 11] = -[1, 22] = [2, 12] = r$$

and the symbols of the second kind are

$$\begin{Bmatrix} 1 \\ 1 \quad 1 \end{Bmatrix} = -\begin{Bmatrix} 1 \\ 2 \quad 2 \end{Bmatrix} = \frac{r}{1 + r^2}, \qquad \begin{Bmatrix} 2 \\ 1 \quad 2 \end{Bmatrix} = \frac{1}{r}.$$

All the other components are zero. Hence the geodesic equations are

$$0 = r'' + \frac{r}{1 + r^2}(r'^2 - \theta'^2)$$

$$0 = \theta'' + \frac{2}{r} r'\theta'.$$

These equations can in principle be solved; the reduction to a single quadrature is straightforward but this final integration is difficult.

One of the most interesting applications of this work is in general relativity and this will be looked at in a little more detail in section 9.3.

7.5* Maximum Enclosed Area

It is at last possible to solve the classical problem posed in example 7. As indicated in chapter 1 the problem is of considerable historical interest. The original proof that the closed curve maximizing the enclosed area is in fact a circle depended on solving the Euler equations. This proof, which is given below, suffers from one major defect. The defect is not restricted to this particular problem but is one which is present in all proofs involving the Euler equation. This equation is deduced on the assumption that an extremum exists and it gives only a *necessary* condition. The question of whether this extremum exists has to be proved separately. The intuitive solution to the maximum area problem was known to the Greeks; nearly two thousand years later the partial solution was established using the Euler equation but a further three hundred years were required before the existence of a solution was proved. The order of magnitude of the difficulty of sufficiency proofs in the calculus of variations is well illustrated by this historical survey.

Returning to the proof of necessity from the Euler equation, suppose that the area R is enclosed by a curve $\mathscr{C}$ then the problem can be stated as maximizing

$$A = \iint_R dx\,dy$$

subject to

$$L = \int_{\mathscr{C}} ds.$$

The major task is to write these in a form for which the standard results can be used. The first step is to parameterize the curve $\mathscr{C}$ as $x = x(t)$, $y = y(t)$ in such a way that as t increases from 0 to T, the arc length along $\mathscr{C}$ increases monotonically and the curve is described once. The length constraint becomes

$$L = \int_0^T (\dot{x}^2 + \dot{y}^2)^{\frac{1}{2}}\,dt, \tag{7.14}$$

while A is computed using the same technique described in example 44.

$$A = \iint_R dx\,dy = \tfrac{1}{2}\oint_{\mathscr{C}} (-y\,dx + x\,dy) = \tfrac{1}{2}\int_0^T (-y\dot{x} + x\dot{y})\,dt. \tag{7.15}$$

Since the problem is independent of the coordinate system it may be assumed without loss of generality that $\mathscr{C}$ passes through the origin and hence the end conditions become

$$x(0) = x(T) = y(0) = y(T) = 0.$$

Maximizing (7.15) subject to (7.14) is equivalent to finding the unconstrained extremum of

$$A^*[x, y] = \tfrac{1}{2}\int_0^T [(-y\dot{x} + x\dot{y}) + \lambda(\dot{x}^2 + \dot{y}^2)^{\frac{1}{2}}]\,dt,$$

where λ is a Lagrange multiplier. The Euler equations are

$$0 = \frac{d}{dt}\left(-y + \frac{\lambda\dot{x}}{(\dot{x}^2 + \dot{y}^2)^{\frac{1}{2}}}\right) - \dot{y}$$

$$0 = \frac{d}{dt}\left(x + \frac{\lambda\dot{y}}{(\dot{x}^2 + \dot{y}^2)^{\frac{1}{2}}}\right) + \dot{x},$$

which can be integrated immediately to give

$$A = -2y + \frac{\lambda\dot{x}}{(\dot{x}^2 + \dot{y}^2)^{\frac{1}{2}}}, \qquad B = 2x + \frac{\lambda\dot{y}}{(\dot{x}^2 + \dot{y}^2)^{\frac{1}{2}}}.$$

From these two equations

$$(A + 2y)^2 + (B - 2x)^2 = \frac{\lambda^2(\dot{x}^2 + \dot{y}^2)}{\dot{x}^2 + \dot{y}^2} = \lambda^2,$$

which is just a *circle* with centre $(\frac{1}{2}B, -\frac{1}{2}A)$ and radius $\frac{1}{2}\lambda$. Note that again the Lagrange multiplier λ is a parameter of some interest since $L = \pi\lambda$.

PROBLEMS

1. Modify the brachistochrone problem in section 7.1 (i) to include the case when the particle has an initial velocity V, (ii) to include the case when $y(a)$ is unspecified (i.e. the bead slides down to any point on the line $x = a$).
2. (Newton's problem.) A rocket nose, in the form of a surface of revolution, is obtained by rotating the curve $y(x)$ between $x = 0$ and $x = a$ about the x-axis. Due to a large number of particles, all moving parallel to the x-axis with constant velocity V and colliding elastically with the nose, it experiences a resistance proportional to

$$\int_0^a \frac{yy'^3}{1 + y'^2}\,dx.$$

$y(0) = 0$, $y(a) = b$. Deduce this result and then show that the minimum resistance is experienced by the curve

$$y = \frac{C(1 + p^2)^2}{p^3}, \qquad x = C\left(\log p + \frac{1}{p^2} + \frac{3}{4}p^4\right) + K,$$

where C and K are arbitrary constants. (N.B. $p = dy/dx$.)

3. Deduce that in a one dimensional problem with a potential energy $V(x)$, an equilibrium point is obtained if $dV/dx = 0$ and $d^2V/dx^2 > 0$ (i.e. V is a minimum). Hence show that a bead at the bottom of a smooth vertical circular wire is stable but at the top is unstable.

4. A uniform rod AB, of mass M and length $2a$, is smoothly hinged to a vertical wall at A. The end B is attached by an elastic string, of natural length a and modulus λ, to a point C immediately above A. Find the equilibrium position of the rod.

5. Discuss the stability of the constant coefficient differential equations

$$\dot{\xi}_i = \sum_{k=1}^{2} a_{ik}\xi_k \qquad (i = 1, 2)$$

at the point $\xi_i = 0$ $(i = 1, 2)$. Consider solutions of the form $\xi_i = A_i \exp(\mu t)$.

6. (Kelvin's theorem.) If $\mathbf{q} = -\nabla\varphi$ and div $\mathbf{q} = 0$ and T is defined by

$$T = \tfrac{1}{2}\rho \int_V \mathbf{q}^2 \, dV$$

show that

$$T = \tfrac{1}{2}\rho \int_S \varphi \frac{\partial\varphi}{\partial n} \, dS.$$

Deduce that if $\partial\varphi/\partial n = 0$ on S then $\mathbf{q} \equiv 0$ in V. If also $\mathbf{q}_1 = -\nabla\varphi + \mathbf{q}_0$, with div $\mathbf{q}_0 = 0$ and $\mathbf{q}_0 \cdot \mathbf{n} = 0$ on S, show that

$$T \leq \tfrac{1}{2}\rho \int \mathbf{q}_1{}^2 \, dV.$$

These results have a standard interpretation in a fluid flow situation.

7. In the statistical mechanical theory of magnetism spins, N_1 'up' and N_2 'down', are distributed on N lattice sites. The interaction between opposite spins is J and $-J$ between like spins. A magnetic field H is applied parallel to the 'up' direction at a temperature T. The mean field approximation leads to a free energy

$$F = -\tfrac{1}{2}qJ(x_1 - x_2)^2 + mH(x_1 - x_2) + kT(x_1 \log x_1 + x_2 \log x_2),$$

where $x_1 = N_1/N$, $x_2 = N_2/N$ and hence $x_1 + x_2 = 1$. The other parameters are all constants. Minimize F with respect to $R = x_1 - x_2$ to deduce the equilibrium condition

$$R = \tanh\left(\frac{qJR - mH}{kT}\right).$$

When $H = 0$ look at the solutions for R in the cases $(kT/qJ) \gtrless 1$.

8. Use Hamilton's principle, section 7.3, to derive the normal equations of motion of a simple pendulum.

9. Deduce that Lagrange's equations (7.8) can be written in Hamiltonian form, with $H(q_1, \ldots, q_N, p_1, \ldots, p_N) = T + V$, as

$$\dot{q}_i = \frac{\partial H}{\partial p_i}, \qquad \dot{p}_i = -\frac{\partial H}{\partial q_i}.$$

This is a non-trivial deduction so consult a book on advanced mechanics in the event of difficulty.

10. Obtain the Christoffel symbols and hence the geodesic equations (7.13) for (i) a sphere ($ds^2 = a^2\, d\theta^2 + a^2 \sin^2\theta\, d\varphi^2$), (ii) a cone, $x^2 + y^2 = z^2 \tan^2 \alpha$.

11. The curve $y = y(x)$ joining the points $(0, \varepsilon)$ and (a, b) is rotated about the x-axis. Show that the surface area of the volume generated is

$$2\pi \int_0^a y(1 + y'^2)^{\frac{1}{2}}\, dx.$$

Find the curve that gives the minimum surface area. What happens when $\varepsilon \to 0$?

Chapter VIII*

Differential Equations and Eigenvalues

8.1* Adjoint Operators

8.1.1* DEFINITION AND PROPERTIES

In dealing with differential equations it is not always necessary to deal with specific equations and it is often somewhat easier to develop a theory in terms of general differential operators, L. In abstract terms an operator is a mapping from one function space to another, while in practical terms it can be regarded just as a shorthand notation for writing a differential expression. For example

$$Lu = \frac{du}{dx} \qquad \text{or} \qquad L \equiv \frac{d}{dx}$$

$$Lu = \frac{\partial^2 u}{\partial x^2} + \frac{\partial^2 u}{\partial y^2} \qquad \text{or} \qquad L \equiv \frac{\partial^2}{\partial x^2} + \frac{\partial^2}{\partial y^2}$$

$$Lu = \frac{d^2 u}{dx^2} + \left(\frac{du}{dx}\right)^2.$$

In this last illustration it is not quite so obvious how to write L explicitly since the notation used for the previous two becomes imprecise. The operator, however, is perfectly well defined.

To produce meaningful answers operators must operate on some specific function or functions. However, one of the main interests in the theory of operators is to find out the rules that they satisfy and how they can be manipulated without reference to any specific function. For instance one particularly important property of an operator is whether or not it is linear. The linearity of L means that

$$L(au + bv) = aLu + bLv$$

for all constants a, b and all functions u, v belonging to a suitably defined class. In the above illustrations it may be noted that the first two are linear while the third is not. For linear operators the important property is that brackets can be multiplied out by the usual laws of algebra. In particular this implies that if $Lu = 0$ and $Lv = 0$ then $L(au + bv) = 0$ so that solutions can be superposed. The present chapter will be devoted almost entirely to linear operators since they lead to very amenable mathematics. As soon as non-linear operators are considered, superposition is lost and final computations are usually very complicated, any variational aspects will almost certainly require the hill climbing techniques of chapters 2 and 3.

In the theory of linear differential operators the adjoint operator is a most important idea since it opens up a whole range of interesting problems and new techniques. For the present purposes this idea will be introduced in the case of a single variable and then extensions to a larger number of variables will be quoted. Let L be a differential operator of a single variable, then the *adjoint operator* M is defined so that for *any* two suitably differentiable functions

$$vLu - uMv = \frac{d}{dx}(\text{function}), \tag{8.1}$$

where the function contains only differentials of u and v up to one order less than that of L or M.

Example 51 Find the adjoint operator of

$$L \equiv \frac{d^2}{dx^2} + \frac{d}{dx} + 1.$$

Consider

$$M \equiv \frac{d^2}{dx^2} - \frac{d}{dx} + 1$$

$$\begin{aligned} vLu - uMv &= vu'' + vu' + uv - uv'' + uv' - uv \\ &= \frac{d}{dx}(vu' - uv' + uv). \end{aligned}$$

Thus M is the adjoint of L.

For the general second order linear operator of a single variable

$$Lu \equiv a(x)u'' + b(x)u' + c(x)u \tag{8.2}$$

the adjoint can be constructed explicitly since

$$\begin{aligned} vLu &= avu'' + bvu' + cuv \\ &= (avu')' - (av)'u' + (bvu)' - (bv)'u + cuv \\ &= u[(av)'' - (bv)' + cv] - [(av)'u - avu' - bvu]' \end{aligned}$$

and hence

$$vLu - uMv = [a(vu' - uv') + (b - a')uv]',$$

where the adjoint M is given by

$$Mv \equiv (av)'' - (bv)' + cv.$$

Such operators are called *self-adjoint* if $L \equiv M$; in this particular case it implies that $a' = b$ and hence the general self-adjoint operator of the form (8.2) is

$$Lu \equiv \frac{d}{dx}(au') + cu. \tag{8.3}$$

Associated with these operators there are normally some boundary conditions. Particularly important amongst these are *adjoint boundary conditions* which are chosen to ensure that for all u, v

$$\int_\alpha^\beta uLv\, dx = \int_\alpha^\beta vMu\, dx \tag{8.4}$$

for the interval $\alpha \leq x \leq \beta$. In particular this implies that

$$\begin{aligned}\int_\alpha^\beta (uLv - vMu)\, dx &= \int_\alpha^\beta \frac{d}{dx}[a(vu' - uv') + (b - a')uv]\, dx \\ &= [a(vu' - uv') + (b - a')uv]_\alpha^\beta\end{aligned}$$

and are satisfied if these conditions $[a(vu' - uv') + (b - a')uv] = 0$ at $x = \alpha$ and $x = \beta$. It may be noted that the common conditions $u = 0$ and $v = 0$ at the end points satisfy these conditions.

There can be some confusion over the definition of adjoint operators since many authors take the definition as (8.1) while others as (8.4) and hence incorporate the boundary conditions in the definition. A little care is needed in this but the definition is usually given by the context. For operators of the type (8.2) it is always possible to convert it to the self-adjoint form (8.3) by multiplying by an appropriate factor (see problem 2). In consequence it is only necessary to consider self-adjoint operators. For partial differential operators this is not necessarily the case and the position is not so straightforward. It is, however, fortunate that many of the important operators are self-adjoint and in particular two of the three important operators of mathematical physics are self-adjoint.

Consider the corresponding definitions for higher dimensions. The partial differential operator L has an adjoint M if for *any* two sufficiently differentiable functions u and v

$$vLu - uMv = \text{div}\,\mathbf{A}, \tag{8.5}$$

where **A** is a vector function of u and v and derivatives of u and v up to one order less than L or M. Again self-adjointness is given if $L \equiv M$.

Example 52 Test for self-adjointness of the operators

$$\text{(a)}\quad L \equiv \frac{\partial^2}{\partial x^2} + \frac{\partial^2}{\partial y^2}, \qquad \text{(b)}\quad M \equiv \frac{\partial^2}{\partial x^2} - \frac{\partial}{\partial t}, \qquad \text{(c)}\quad N \equiv \frac{\partial^2}{\partial x^2} - \frac{\partial^2}{\partial t^2}.$$

$$\text{(a)}\quad vLu - uLv = \frac{\partial}{\partial x}(vu_x - uv_x) + \frac{\partial}{\partial y}(vu_y - uv_y)$$

and the Laplace operator is therefore self-adjoint.

$$\text{(b)}\quad vMu - uMv = \frac{\partial}{\partial x}(vu_x - uv_x) - vu_t + uv_t$$

which is not self-adjoint. The heat conduction operator M has an adjoint $(\partial^2/\partial x^2 + \partial/\partial t)$ which can be checked by the reader.

$$\text{(c)}\quad vNu - uNv = \frac{\partial}{\partial x}(vu_x - uv_x) + \frac{\partial}{\partial t}(uv_t - vu_t)$$

and the wave operator is self-adjoint.

In parallel with the above analysis, for the general second order partial linear differential operator

$$Lu \equiv \sum_{i,j} a_{ij} \frac{\partial^2 u}{\partial x_i\, \partial x_j} + \sum_i b_i \frac{\partial u}{\partial x_i} + cu \tag{8.6}$$

its adjoint can be shown to be

$$Mv \equiv \sum_{i,j} \frac{\partial^2 (a_{ij} v)}{\partial x_i\, \partial x_j} - \sum_i \frac{\partial (b_i v)}{\partial x_i} + cv \tag{8.7}$$

since it can be verified by the reader that (it is assumed without loss of generality that $a_{ij} = a_{ji}$)

$$vLu - uMv = \sum_i \frac{\partial}{\partial x_i}\left[\left(b_i - \sum_j \frac{\partial a_{ij}}{\partial x_j}\right) uv + \sum_j a_{ij}\left(v\frac{\partial u}{\partial x_j} - u\frac{\partial v}{\partial x_j}\right)\right].$$

To be self-adjoint ($L \equiv M$) the condition is that

$$b_i = \sum_j \frac{\partial a_{ij}}{\partial x_j}$$

and the self-adjoint boundary conditions for a region R with boundary S are obtained from

$$\int_R (vLu - uLv)\, dV = \int_R \sum_i \frac{\partial}{\partial x_i} \sum_j a_{ij}\left(v\frac{\partial u}{\partial x_j} - u\frac{\partial v}{\partial x_j}\right) dV$$

$$= \int_S \sum_i \sum_j a_{ij}\left(v\frac{\partial u}{\partial x_j} - u\frac{\partial v}{\partial x_j}\right) n_i\, dS,$$

where the general divergence theorem has been used and $\hat{\mathbf{n}} = (n_1, n_2, \ldots)$ is the unit normal to S. The vanishing of the integrand of the right-hand side of this expression provides these self-adjoint conditions.

8.1.2* VARIATIONAL PROPERTIES

One of the important aspects of the adjoint can be seen from considering some variational problems concerned with the linear differential equation

$$Lu = 0 \quad \text{in } R, \tag{8.8}$$

where it is assumed that adjoint boundary conditions are satisfied so that $\int uLw\, dV = \int wMu\, dV$. For the extremum of the functional

$$J[u, w] = \int_R wLu\, dV \tag{8.9}$$

consider the usual variations $u = U + \varepsilon\eta$ and $w = W + \varepsilon\xi$ giving

$$J[u, w] = J[U, W] + \varepsilon\int_R (\xi LU + WL\eta)\, dV + 0(\varepsilon^2).$$

Applying the argument that $\varepsilon = 0$ gives an extremum of this expression then

$$0 = \int_R (\xi LU + WL\eta)\, dV = \int_R (\xi LU + \eta MW)\, dV.$$

Now the ξ and η are arbitrary except for satisfying zero boundary conditions and hence the Euler equations become

$$LU = 0, \qquad MW = 0.$$

Thus finding the extremum of (8.9) gives a solution of (8.8). A particularly important special case is the self-adjoint problem where the functional

$$I[u] = \int_R uLu\, dV \tag{8.10}$$

gives the same desired result so that the Euler equation is

$$Lu = 0.$$

Example 53 Deduce that $L \equiv \nabla^2$ is self-adjoint and find the corresponding functional.

The identity

$$w\,\nabla^2 u - u\,\nabla^2 w = \operatorname{div}(w\,\nabla u - u\,\nabla w)$$

shows immediately that the operator is self-adjoint, so that

$$\int_R (w\,\nabla^2 u - u\,\nabla^2 w)\,dV = \int_S (w\,\nabla u - u\,\nabla w)\,.\,\hat{\mathbf{n}}\,dS$$

and hence self-adjoint conditions must give $(w\,\partial u/\partial n - u\,\partial w/\partial n) = 0$ on the boundary S. The corresponding functional is

$$I[u] = \int_R u\,\nabla^2 u\,dV = \int_R [\operatorname{div}(u\,\nabla u) - (\nabla u)^2]\,dV$$

$$= -\int_R (\nabla u)^2\,dV$$

provided $u\,\partial u/dn = 0$ on S. This last form is probably the most convenient to use.

Example 54 Deduce the Euler equation for the functional

$$I[u, w] = \int_R (wLu - \tfrac{1}{2}w^2)\,dV$$

with suitable adjoint boundary conditions satisfied.

Let $u = U + \varepsilon\eta$, $w = W + \varepsilon\xi$ in the usual way and apply the standard arguments without explanation

$$I[u, w] = I[U, W] + \varepsilon \int_R (WL\eta + \xi LU - W\xi)\,dV + 0(\varepsilon^2).$$

The extremum is therefore given by

$$0 = \int_R (WL\eta + \xi LU - W\xi)\,dV = \int_R [\eta MW + \xi(LU - W)]\,dV,$$

where M is the adjoint of L. Thus since η and ξ are arbitrary

$$MW = 0, \qquad LU - W = 0$$

or

$$(ML)U = 0.$$

A whole variety of similar variational formulae can be deduced from different functionals and some of these will be developed later in this chapter.

8.1.3* EIGENVALUES

Given a linear self-adjoint operator L great interest is centred around the non-trivial solutions of the eigenvalue equation

$$Lu = \lambda u \quad \text{in } R, \tag{8.11}$$

where λ is a constant and $u = 0$ on the boundary S of R.

Example 55 Find the values of λ for which non-trivial solutions exist for

$$u'' + \lambda^2 u = 0$$

in $a \leq x \leq b$ with $u(a) = u(b) = 0$.

Independent solutions of this differential equation are $e^{i\lambda x}$, $e^{-i\lambda x}$ and hence the general solution is

$$u = Ae^{i\lambda x} + Be^{-i\lambda x}.$$

The boundary conditions imply that

$$u(a) = 0 = Ae^{i\lambda a} + Be^{-i\lambda a}$$
$$u(b) = 0 = Ae^{i\lambda b} + Be^{-i\lambda b}$$

and a non-trivial solution exists for A and B only if

$$0 = e^{i\lambda(b-a)} - e^{-i\lambda(b-a)} = 2i \sin \lambda(b - a),$$

and hence only if $\lambda = N\pi/(b - a)$, where N is an integer.

It is clear from this example that the eigenvalue problem cannot be solved for all λ but only for particular numbers or *eigenvalues*. These have considerable importance both mathematically and physically. For instance, the normal modes of decay of the diffusion equation (in two dimensions)

$$\frac{\partial^2 u}{\partial x^2} + \frac{\partial^2 u}{\partial y^2} = \frac{\partial u}{\partial t}$$

can be deduced by putting $u = e^{-\lambda t}v(x, y)$, whence

$$\nabla^2 v + \lambda v = 0.$$

Similarly the wave equation describing the motion of sound waves is

$$\frac{\partial^2 u}{\partial x^2} + \frac{\partial^2 u}{\partial y^2} + \frac{\partial^2 u}{\partial z^2} = \frac{1}{c^2}\frac{\partial^2 u}{\partial t^2},$$

and putting periodic solutions $u = e^{i\omega t}v(x, y, y)$ gives

$$\nabla^2 v + \left(\frac{\omega}{c}\right)^2 v = 0,$$

and the eigenvalues give the note or frequency of the sound wave.

Some attempt at the eigenvalue problem was made in example 40 in section 6.4.1 and the present analysis generalizes the example for a general self-adjoint linear operator L. The problem is the isoperimetric one of finding the extremum

$$I[u] = \int_R uLu\, dV \tag{8.12}$$

subject to

$$1 = \int_R u^2\, dV \tag{8.13}$$

and self-adjoint conditions on the boundary S of R. Using a Lagrange multiplier λ the unconstrained extremum of

$$I^*[u] = \int_R (uLu - \lambda u^2)\, dV$$

is therefore required. Working in the usual way with $u = U + \varepsilon\eta$ together with the self-adjoint boundary conditions gives, with a little standard working,

$$LU = \lambda U.$$

Now when this extremum is substituted into (8.12)

$$I[U] = \int_R ULU\, dV = \int_R U\lambda U\, dV$$

$$= \lambda \int_R U^2\, dV = \lambda \qquad \text{(from 8.13)}.$$

Hence $I[U]$ provides the value of λ. This is in fact the lowest eigenvalue of (8.11). To compute the higher values further restrictions must be placed in addition to (8.13) and these are obtained from a general result applicable to self-adjoint operators.

Suppose

$$Lu_1 = \lambda_1 u_1 \quad \text{and} \quad Lu_2 = \lambda_2 u_2$$

and $\lambda_1 \neq \lambda_2$ then since L is self-adjoint

$$0 = \int_R (u_2 Lu_1 - u_1 Lu_2)\, dV = \int_R (\lambda_1 u_1 u_2 - \lambda_2 u_1 u_2)\, dV$$

$$= (\lambda_1 - \lambda_2) \int_R u_1 u_2\, dV.$$

Thus since $\lambda_1 \neq \lambda_2$ then

$$\int_R u_1 u_2 \, dV = 0 \tag{8.14}$$

so that the eigenfunctions corresponding to different eigenvalues are orthogonal in the sense (8.14).

Taking the lead from this result, suppose $u_1, u_2, \ldots, u_n$ are the first n normalized eigenfunctions (i.e. $\int_R u_i^2 \, dV = 1$) corresponding to the n lowest eigenvalues $\lambda_1, \lambda_2, \ldots, \lambda_n$ ($\lambda_1 \leq \lambda_2 \leq \cdots \leq \lambda_n$) and assume that these are all known. Find the extremum of

$$I[u] = \int_R uLu \, dV$$

subject to self-adjoint conditions on S the boundary of R and

$$1 = \int_R u^2 \, dV$$

and

$$0 = \int_R uu_i \, dV, \qquad i = 1, \ldots, n. \tag{8.15}$$

Introducing Lagrange multipliers $\lambda, \mu_1, \mu_2, \ldots, \mu_n$, the modified functional takes the form

$$I^{**}[u] = \int_R \left(uLu - \lambda u^2 - 2 \sum \mu_i uu_i \right) dV$$

with Euler equation for the extremum U as

$$LU = \lambda U + \sum \mu_i u_i. \tag{8.16}$$

Again

$$\begin{aligned} I[U] &= \int ULU \, dV = \lambda \int U^2 \, dV + 2 \sum \mu_i \int Uu_i \, dV \\ &= \lambda \end{aligned}$$

using the normalization and (8.15). The values of μ_i at the extremum can be computed by multiplying (8.16) by u_j and integrating over the whole of R.

$$\int_R u_j LU \, dV = \int_R \lambda u_j U \, dV + \sum \mu_i \int_R u_j u_i \, dV$$

or

$$\int_R ULu_j \, dV = \mu_j$$

or

$$\mu_j = \int_R \lambda_j U u_j \, dV = 0.$$

Thus at the extremum the μ_j are zero, (8.16) shows that the eigenvalue equation $LU = \lambda U$ is satisfied, and $I[U] = \lambda$ gives the next eigenvalue.

Having constructed the variational formulation the whole power of Rayleigh–Ritz technique can be used on the detailed computation of the eigenvalues of an operator.

8.1.4* INSTABILITY

Take one particular example to illustrate how the above results can be used to decide whether a system is unstable. Consider the equation

$$Lu = \frac{\partial^2 u}{\partial t^2},$$

where L is a linear self-adjoint operator in the x, y, z variables. Periodic solutions of the form $u = e^{i\omega t}\psi(x, y, z)$ give

$$L\psi + \omega^2\psi = 0 \tag{8.17}$$

and this will be said to have unstable modes if a non-trivial solution can be found with $\omega^2 < 0$. This would imply that there exist exponentially increasing solutions. It is assumed that besides the orthogonality condition (8.14) the normalized eigenfunctions $\psi_1, \psi_2, \ldots, \psi_n, \ldots$ form a complete set (cf. section 5.2) so that any function in the space can be expanded in terms of them as accurately as required.

If there exists a function φ satisfying zero boundary conditions so that

$$\int_R \varphi L\varphi \, dV > 0$$

then (8.17) has unstable modes. This can be proved by writing

$$\varphi = \sum_{i=1}^{\infty} a_i\psi_i$$

$$\begin{aligned} 0 < \int_R \varphi L\varphi \, dV &= \int_R \sum_{i,j} a_i a_j \psi_i L\psi_j \, dV \\ &= \sum_{i,j} a_i a_j \int_R (-\omega_j^2)\psi_i\psi_j \, dV \\ &= -\sum_{i=1}^{\infty} a_i^2\omega_i^2. \end{aligned}$$

Since a_i are real constants the only alternative is that at least one of the $(\omega_i^2) < 0$ and hence instability.

Example 56 Show that the differential equation

$$\frac{\partial^2 u}{\partial x^2} + 40x^2 u = \frac{\partial^2 u}{\partial t^2}$$

has unstable modes; boundary conditions are taken as $u(0, t) = u(1, t) = 0$ for all t.

The operator $L \equiv (d^2/dx^2) + 40x^2$ is shown easily to be self-adjoint and try the function $\varphi = x(1 - x)$ which satisfies the boundary conditions.

$$\int_0^1 \varphi L\varphi \, dx = \int_0^1 (-2x + 2x^2 + 40x^4 - 80x^5 + 40x^6) \, dx$$

$$= \frac{1}{21} > 0$$

and hence instability.

8.2* RAYLEIGH–RITZ

8.2.1* GENERAL APPLICATION

The direct method of Rayleigh–Ritz gives a powerful tool when applied to the eigenvalue equations (8.12), (8.13), (8.15). It will be recalled that the procedure selects as a trial function, a linear combination of a finite number of functions belonging to a complete set $\varphi_1, \varphi_2, \ldots, \varphi_n, \ldots$ which satisfy appropriate boundary conditions, in the present case zero boundary conditions. The expansion involving the first N functions is

$$u = a_1\varphi_1 + a_2\varphi_2 + \cdots + a_N\varphi_N. \tag{8.18}$$

For the lowest eigenvalue substitute into (8.12) and (8.13) to give the expression

$$I[u] = \sum_{i,j} a_i a_j \int_R \varphi_i L\varphi_j \, dV$$

subject to

$$1 = \sum_{i,j} a_i a_j \int_R \varphi_i \varphi_j \, dV.$$

Since the φ_i are known, the constants

$$L_{ij} = L_{ji} = \int_R \varphi_i L\varphi_j \, dV, \qquad S_{ij} = S_{ji} = \int_R \varphi_i \varphi_j \, dV$$

can be computed (the symmetry $L_{ij} = L_{ji}$ is obtained from the self-adjointness of the operator L) by simple quadrature. Thus the problem is to find the extremum of

$$I[u] = \sum a_i a_j L_{ij}$$

subject to

$$1 = \sum a_i a_j S_{ij}.$$

The constraint can be removed by using a Lagrange multiplier λ to obtain the modified functional

$$I^*[u] = \sum a_i a_j (L_{ij} - \lambda S_{ij}).$$

The extremum with respect to the a_j is just

$$\frac{\partial I^*}{\partial a_j} = 2 \sum_i (L_{ij} - \lambda S_{ij}) a_i = 0, \quad j = 1, 2, \ldots, N, \tag{8.19}$$

which is a matrix equation for a_i having a non-trivial solution if the determinant

$$\begin{vmatrix} L_{11} - \lambda S_{11} & L_{12} - \lambda S_{12} & \cdots & L_{1N} - \lambda S_{1N} \\ L_{21} - \lambda S_{21} & L_{22} - \lambda S_{22} & & \cdot \\ \cdot & \cdot & & \cdot \\ \cdot & \cdot & & \cdot \\ \cdot & \cdot & & \cdot \\ L_{N1} - \lambda S_{N1} & \cdots & & L_{NN} - \lambda S_{NN} \end{vmatrix} = 0. \tag{8.20}$$

This determinant is a polynomial of degree N in λ and hence has N roots $\lambda_1, \ldots, \lambda_N$ and for each root the equations (8.19) can be solved to give a corresponding set of a's $a_1^{(1)}, a_2^{(1)}, \ldots, a_N^{(1)}$; $a_1^{(2)}, \ldots, a_N^{(2)}; \ldots;$ $a_1^{(N)}, a_2^{(N)}, \ldots, a_N^{(N)}$. The corresponding functions (8.18) are

$$u_k = a_1^{(k)}\varphi_1 + a_2^{(k)}\varphi_2 + \cdots + a_N^{(k)}\varphi_N \tag{8.21}$$

and these have the properties of approximate eigenfunctions of the equation $Lu = \lambda u$. This can be seen by first computing

$$\begin{aligned} \int u_m L u_k \, dV &= \sum_{i,j} L_{ij} a_i^{(k)} a_j^{(m)} \\ &= \sum_{i,j} \lambda_k S_{ij} a_i^{(k)} a_j^{(m)} \qquad \text{(from (8.19))} \\ &= \lambda_k \int u_m u_k \, dV. \end{aligned}$$

The last step comes from the use of the definition of the Ss and (8.21). Thus if $m = k$, because of the normalization,

$$\lambda_k = \int u_k L u_k \, dV,$$

while if $m \neq k$ because of the self-adjointness of L

$$\int u_k L u_m \, dV = \lambda_m \int u_m u_k \, dV$$

and

$$(\lambda_k - \lambda_m) \int u_m u_k \, dV = 0.$$

Therefore if $\lambda_k \neq \lambda_m$ the functions u_m, u_k are orthogonal. Comparing this analysis with the analysis of section 8.1.3 it can be seen that the roots of (8.20) and the corresponding functions (8.21) give approximations to the first N eigenvalues and eigenfunctions of

$$Lu = \lambda u.$$

The case of degenerate eigenvalues takes a bit more care than the deduction presented here but the same results follow through (see Courant and Hilbert). This reference also deals with the rigorous convergence of the method.

As a final note to this section, the normalization (8.13) is often dispensed with and the functional

$$J[u] = \frac{\int uLu \, dV}{\int u^2 \, dV}$$

is considered. This is called the Rayleigh quotient and it is left to the reader to show that equivalent results hold.

8.2.2* STURM–LIOUVILLE PROBLEM

A particularly suitable application of the above work considers the following problem, studied originally by Sturm and Liouville.

Find solutions $u(x) \not\equiv 0$ in the closed (finite) interval $a \leq x \leq b$ of the differential equation

$$[r(x)u']' + p(x)u + \lambda q(x)u = 0 \tag{8.22}$$

with

$$u(a) = u(b) = 0;$$

the conditions on the functions p, q and r are that $r(x) > 0$ and $q(x) > 0$ in $a \leq x \leq b$ and that they are all differentiable functions of x in the interval.

The advantage of this system is that the operator is self-adjoint, see (8.3), and it can be shown quite generally (see Courant and Hilbert (1953)) that there is an infinite sequence of eigenvalues $\lambda_1, \lambda_2, \ldots$ of (8.22) with the properties

$$\lambda_1 < \lambda_2 < \lambda_3 < \cdots, \qquad \lambda_n \to \infty \quad \text{as } n \to \infty.$$

The corresponding sequence of eigenfunction solutions $u_1(x), u_2(x), \ldots$ is complete and has the properties that $u_n(x)$ has n zeros in $a < x < b$ and the slightly extended orthogonality conditions hold

$$\int_a^b q(x)u_i(x)u_j(x)\,dx = 0 \quad \text{if } i \neq j.$$

Consequent on this theorem is the comforting result that discrete non-degenerate eigenvalues exist and it is only a matter of finding a satisfactory method of extracting them numerically. The Rayleigh–Ritz method provides just such a method.

Example 57 Find approximations to the first three eigenvalues of

$$-u'' + x^2 u = \lambda u$$

with $u = 0$ at $x = \pm 1$.

Firstly it should be noted that the operator is self-adjoint and it is of the Sturm–Liouville type (8.22). Secondly it is necessary to find a set of approximating functions; these can be chosen as

$$\varphi_1 = (1 - x^2) \qquad \varphi_2 = x(1 - x^2), \qquad \varphi_3 = x^2(1 - x^2).$$

Thirdly and finally the detailed computations following section 8.2.1 must be performed.

After some tedious but trivial integrations it can be found that

$$L_{11} = 2{\cdot}819\,0,\ L_{12} = 0,\ L_{13} = 0.584\,1,\ L_{22} = 1.650\,8,\ L_{32} = 0,$$
$$L_{33} = 0.861\,2$$

and

$$S_{11} = 1.066\,7,\ S_{12} = 0,\ S_{13} = 0.152\,4,\ S_{22} = 0.152\,4,\ S_{32} = 0,$$
$$S_{33} = 0.050\,8.$$

The first approximation including only the function φ_1 gives an estimate of the lowest eigenvalue as

$$\lambda_1 = \frac{L_{11}}{S_{11}} = 2.643.$$

The second approximation involving φ_1 and φ_2 is found by solving

$$\begin{vmatrix} 2.819\,0 - 1.066\,7\lambda & 0 \\ 0 & 1.650\,8 - 0.152\,4\lambda \end{vmatrix} = 0$$

or

$$\lambda_1 = 2.643, \qquad \lambda_2 = 10.83.$$

The third approximation is found from the solution of the determinantal equation

$$\begin{vmatrix} 2.819\,0 - 1.066\,7\lambda & 0 & 0.584\,1 - 0.152\,4\lambda \\ 0 & 1.650\,8 - 0.152\,4\lambda & 0 \\ 0.584\,1 - 0.152\,4\lambda & 0 & 0.861\,2 - 0.050\,8\lambda \end{vmatrix} = 0$$

giving $\lambda_1 = 2.597$, $\lambda_2 = 10.83$, $\lambda_3 = 25.95$.

At the expense of more integrations, further approximations can be found to successively improve these values.

The Sturm–Liouville type is a particularly important class of differential equations for which the above method works extremely well. It is unfortunate, however, that the special functions of mathematical physics do not fall into this category. For instance, the Bessel and Legendre equations do not fall into the class (8.22) since $r(x)$ is not strictly positive having a zero at the end points. The proof of the Sturm–Liouville theorem fails because of the possibility of solutions containing infinities; despite this, similar results can indeed be obtained. The Hermite and Laguerre equations do not belong to the class (8.22) for a different reason, namely that the interval of definition is infinite. Extra difficulties are thus included but again the theorem can be modified accordingly.

8.3* Poisson and Helmholtz Equations

In partial differential equations a particularly important pair, for which approximate procedures are suitable, are the equations

$$\nabla^2 u = f \qquad \text{Poisson equation} \tag{8.23}$$

$$\nabla^2 u + \lambda u = 0 \qquad \text{Helmholtz equation} \tag{8.24}$$

where f is a known function. These include the three most commonly studied equations, the Laplace equation by putting $f = 0$ in (8.23) and the heat conduction and wave equations by the reduction to (8.24) as shown in section 8.1.3. The operator $L \equiv \nabla^2$ occurring in all these equations is self-adjoint (see example 53) and hence the variational methods developed above can be applied.

The particular difficulty of these equations is the enormously wide variety of boundary conditions and trying to choose suitable functionals to match these. The three main types of conditions suitable to these equations in a region R with boundary S are

(1) u given on S — Dirichlet conditions

(2) $\dfrac{\partial u}{\partial n}$ given on S — Neumann conditions

(3) $au + \dfrac{\partial u}{\partial n}$ given on S — Robin conditions.

Looking back again to example 53 it was shown that the functional $\int u \nabla^2 u \, dV$ was a suitable functional for the Laplace equation with self-adjoint boundary conditions. Unfortunately a little more than this is needed since the self-adjoint boundary condition $u \, \partial u/\partial n = 0$ is a bit too restrictive. However, this does give a lead on how to derive the appropriate functionals, particularly as it was shown in example 53 that the alternative functional $\int (\nabla u)^2 \, dV$ could be used for the Laplace equation.

Consider therefore

$$J_1[u] = \int_R [(\nabla u)^2 + 2fu] \, dV \tag{8.25}$$

with u given on S. Applying the Euler technique with $u = U + \varepsilon\eta$

$$J_1[u] = J_1[U] + 2\varepsilon \int_R (\nabla U \, . \, \nabla\eta + f\eta) \, dV + 0(\varepsilon^2)$$

gives on extremization $\partial J_1/\partial\varepsilon = 0$ at $\varepsilon = 0$

$$0 = \int_R (f\eta + \nabla U \, . \, \nabla\eta) \, dV$$

$$= \int_R (f - \nabla^2 U)\eta \, dV + \int_S \eta \, \nabla U \, . \, \mathbf{dS}. \tag{8.26}$$

If u is given on the boundary (Dirichlet condition) then U must satisfy these conditions and hence $\eta = 0$ on S. Thus by the basic lemma

$$\nabla^2 U = f,$$

as required. Alternatively if the value of u is left unspecified on the boundary then the natural boundary conditions $\partial U/\partial n = 0$ on S are satisfied. This is a special case of the Neumann conditions (2). To obtain the general conditions consider the functional

$$J_2[u] = \int_R [(\nabla u)^2 + 2fu]\, dV + 2\int_S gu\, dS. \tag{8.27}$$

The variation $u = U + \varepsilon\eta$ gives by a precisely similar analysis to the above

$$0 = \int_R (f - \nabla^2 U)\eta\, dV + \int_S \eta\left(\frac{\partial U}{\partial n} + g\right) dS$$

which gives the Poisson equation $\nabla^2 U = f$ as the Euler equation and natural boundary conditions

$$\frac{\partial U}{\partial n} + g = 0 \quad \text{on } S.$$

Note that the class of functions for variation must be left perfectly free on the boundary to produce these conditions. Note also that because Neumann conditions involve derivatives alone, the solution is only unique to an arbitrary constant and further the functions f and g are not independent since it is necessary (problem 8) that $\int_R f\, dV = \int_S g\, dS$. Finally the Robin boundary condition (3) can be deduced from

$$J_3[u] = \int_R [(\nabla u)^2 + 2fu]\, dV + \int_S (2gu + au^2)\, dS \tag{8.28}$$

with Euler equation

$$\nabla^2 U = f \quad \text{in } R$$

and natural boundary conditions

$$\frac{\partial U}{\partial n} + aU + g = 0.$$

An example which incorporates these ideas and indeed shows how an extension of the method can deal with a *mixed* boundary value problem (that is different conditions on different parts of the boundary) is the following:

Example 58 Solve the two dimensional Laplace equation $\nabla^2 u = 0$ in the region R bounded by the curves $y = 0$, $x = 0$, $x = 1$ and $y = 1 + x$. The boundary conditions, referring to figure 8.1 are on S_1 and S_3, $\partial u/\partial x = 0$; on S_2, $u = 0$; on S_4, $u = 1$.

The functional

$$I[u] = \int_R (\nabla u)^2\, dV$$

is in fact satisfactory for this problem since from (8.26) u is specified on S_2 and S_4 and therefore $\eta = 0$ and $\partial u/\partial n = 0$ on S_1 and S_3 so satisfying the natural boundary conditions. In this current problem therefore approximations must be sought which satisfy the correct boundary conditions on S_2 and

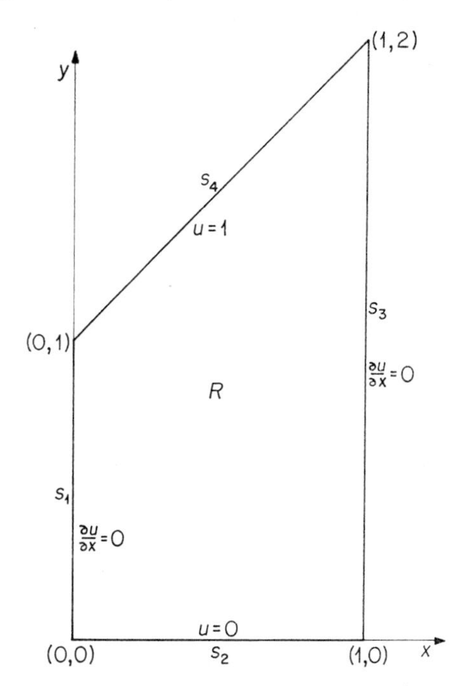

Figure 8.1 Region for solution of Laplace equation, example 58.

S_4 while being left free on S_1 and S_3. A suitable function that may be tried is

$$u = \frac{y}{1 + x}[1 + A(y - 1 - x)].$$

Substituting and performing the integrations gives

$$I[u] = \frac{4}{3} \ln 2 + \frac{1}{2} A + \frac{4}{5} A^2$$

and hence the optimum is given by $A = -\frac{5}{16}$. It may of course be noted that with such a simple approximation the exact natural boundary condition $\partial u/\partial x = 0$ at $x = 0$, $x = 1$ is not satisfied. At $x = 0$ it may be calculated that $\partial u/\partial x = \frac{5}{16}y^2 - y$ and for this approximation it is the best that can be achieved.

For the Helmholtz equation (8.24), the associated eigenvalue problems again usually insist on boundary conditions either $u = 0$ or $\partial u/\partial n = 0$. These are self-adjoint and hence the whole theory in sections 8.1 and 8.2 follows through in a straightforward manner. The only complication is for the appropriate Robin boundary condition (3), i.e. solving

$$\nabla^2\psi + \lambda\psi = 0 \quad \text{in } R$$

subject to

$$\frac{\partial\psi}{\partial n} + a\psi = 0 \quad \text{on } S.$$

Following the usual methods it can be shown that the Rayleigh quotient

$$\frac{(\int_R (\nabla\psi)^2\, dV + \int_S a\psi^2\, dS)}{\int_R \psi^2\, dV}$$

is satisfactory.

8.4* Maximum Principles in Differential Equations

It is of considerable value in the theory of differential equations to produce quite general results for a whole class of problems. Among such results are various maximum principles of which two will be described in detail. For a comprehensive account the reader is referred to the book by Protter and Weinberger (1967). In ordinary differential equations the most important results are those required in the proof of the properties of Sturm–Liouville equations (cf. section 8.2.2).

As a typical example of the sort of maximum principle that can be easily derived consider the following. Let u be a non-trivial solution of

$$u'' + g(x)u' + h(x)u = 0 \tag{8.29}$$

in $a \leq x \leq b$, with the conditions that g and h are bounded and $h(x) \leq 0$ in $a \leq x \leq b$. It can now be stated that u does not have a positive maximum at any interior point c ($a < c < b$) of the interval. The proof of this result is by contradiction and is quite straightforward. Suppose that a positive maximum occurs at $x = c$ then it must be a local maximum with $u(c) > 0$ and $u'(c) = 0$. Now $g(c)$ is bounded so that $gu' = 0$ at $x = c$; also by hypothesis $h(c) < 0$ and $hu < 0$ at $x = c$. But u satisfies (8.29) and hence $u'' > 0$ at $x = c$ which is a contradiction against the hypothesis that a maximum occurs at $x = c$. The result therefore follows that under the given conditions (8.29) has no positive maximum interior to $[a, b]$.

This theorem can be checked for the particular case $u'' - u = 0$, since the exact solution can be computed and the result verified. Application to more complicated equations can be of considerable assistance in assessing the general behaviour of the solution.

Example 59 Show that the solution of

$$u'' - xu' - (\cos \tfrac{1}{2}x)u = 0$$

in $0 \le x \le \frac{1}{2}\pi$, with $u(0) = 1$, $u'(0) = 1$ has no positive maximum in the interval.

The conditions of the above result are satisfied since $(-x)$ and $(-\cos \frac{1}{2}x)$ are bounded in the interval and $(-\cos \frac{1}{2}x) \le 0$ in $0 \le x \le \frac{1}{2}\pi$. Thus the theorem states that the solution has no positive maximum in the interval. Since the initial conditions state that $u(0) > 0$ and u has positive slope at $x = 0$ it may be immediately concluded from the theorem that the solution is *monotonically increasing* in the interval $0 \le x \le \frac{1}{2}\pi$. This is a non-trivial result for a complicated differential equation and has been obtained with very little effort. It is precisely this sort of information that can be extremely valuable both from a physical standpoint and also in the inevitable numerical solution of the differential equation.

Theorems of the above type can be produced with a relaxation of the strict conditions imposed and also for extended versions of the equation. The reader should consult Protter and Weinberger as a suitable source book for these results.

Perhaps the best known maximum principles occur in partial differential equations and in particular for the Laplace equation

$$\nabla^2 u = 0 \quad \text{in } \tau. \tag{8.30}$$

Thinking first of this equation as the steady state heat conduction equation, it is physically reasonable to suppose that the temperature at any point of τ lies between the hottest and coldest temperatures imposed on the boundary. In mathematical terms this implies that the solution of (8.30) has its maximum and minimum values on the boundary of τ.

Consider the situation of solving (8.30) in a sphere, τ_r, of radius r with surface S_r, then by the divergence theorem

$$0 = \int_{\tau_r} \nabla^2 u \, dV = \int_{S_r} \frac{\partial u}{\partial r} \, dS.$$

Using standard spherical polar coordinates this can be rewritten

$$0 = \int_0^{2\pi} \int_0^{\pi} \frac{\partial u}{\partial r} \sin \theta \, d\theta \, d\varphi,$$

and integrating this equation from $r = 0$ to $r = R$ gives

$$0 = \int_0^R \int_0^{2\pi} \int_0^{\pi} \frac{\partial u}{\partial r} \sin\theta \, d\theta \, d\varphi \, dr = \int_0^{2\pi} \int_0^{\pi} [u(R, \theta, \varphi) - u(0, \theta, \varphi)] \sin\theta \, d\theta \, d\varphi.$$

But $u(0, \theta, \varphi)$ is the same whatever the value of θ and φ and is just the value, $u(0)$, of u at the centre of the sphere. Hence

$$u(0) = \frac{1}{4\pi R^2} \int_{S_R} u(R, \theta, \varphi) \, dS$$

so that the value of u at the centre of the sphere τ_R is equal to the average value of u over the surface, S_R, of the sphere.

Now in a region τ enclosed by a smooth surface S, suppose that a non-constant solution of (8.30) can be found. Suppose further that at a point P inside τ a local maximum (a minimum can be treated similarly) occurs, then $u(P) \geq u(Q)$ for all points Q sufficiently close to P. Since P is interior to τ a sphere τ_r can be drawn round P which is entirely in τ, $\tau_r \subset \tau$. Let the sphere τ_r be sufficiently small so that $u(P) \geq u(Q)$ for all points on the surface S_r of τ_r then

$$u(P) = \frac{1}{4\pi r^2} \int_{S_r} u(Q) \, dS \leq \frac{u(P)}{4\pi r^2} \int_{S_r} dS = u(P).$$

Strict equality is obtained only if $u(P) = u(Q)$ for *all* points Q and hence u is constant in the neighbourhood of P and it follows in the whole of τ. But u was assumed to be not a constant and therefore the strict inequality holds, hence $u(P) < u(P)$ and a contradiction is obtained. The result finally proved is that a non-constant solution of (8.30) has no maximum or minimum in the interior of τ.

Two of the major uses of this result are firstly the theoretical one of proving a uniqueness theorem for (8.30) and secondly in practical assistance in numerical solutions. In this latter case it helps in 'debugging' and checking computations or approximation formulae since it is now known that 'hot spots' must not occur and that the average property must hold. The uniqueness theorem mentioned is easily proved for Dirichlet conditions on S, the boundary of a closed finite region τ. Let

$$\nabla^2\psi = 0 \quad \text{in } \tau \qquad \text{with } \psi = f(P) \quad \text{on } S$$

$$\nabla^2\varphi = 0 \quad \text{in } \tau \qquad \text{with } \varphi = f(P) \quad \text{on } S$$

then

$$\nabla^2(\psi - \varphi) = 0 \quad \text{in } \tau \qquad \text{with } (\psi - \varphi) = 0 \quad \text{on } S.$$

But $(\psi - \varphi)$ in τ lies between the maximum and minimum values of $(\psi - \varphi)$ on S, thus $(\psi - \varphi) \equiv 0$ in τ and uniqueness follows.

These results can of course be extended to cover other boundary conditions, unbounded regions, general elliptic linear operators and in certain cases non-linear operators. For other types of equation, heat conduction and wave equation, for example, maximum principles can be established and the reader is referred to problem 11 and the book already mentioned for further details.

8.5* Finite Element Methods in Numerical Analysis

A modern development in the solution of elliptic (i.e. of Laplace type) partial differential equations has been finite element methods. These originated in structural analysis and much of the notation in the literature is taken from civil engineering. In the mathematical context given here a variational approach is used to obtain the basic set of linear algebraic equations; these are then solved by any standard procedure to obtain the numerical solution. Finite elements were first introduced to overcome the difficulties of ordinary

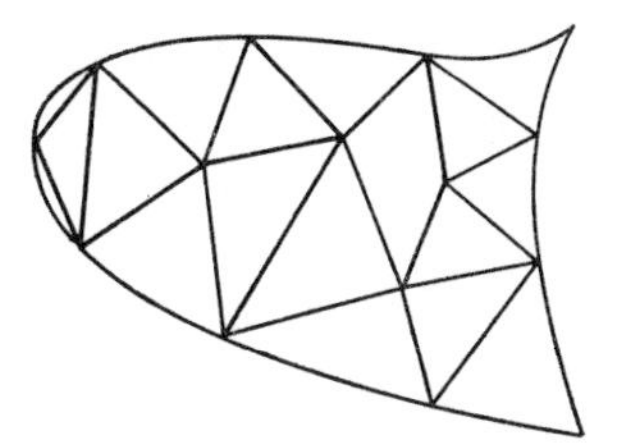

Figure 8.2 Irregular triangular mesh.

finite difference methods when dealing with awkward boundaries. Such boundaries occur, of course, with monotonous regularity in engineering problems. To deal with these boundaries an irregular triangular mesh is normally used in finite element methods and even the most awkward boundaries can be approximated very well by a set of line segments as illustrated in figure 8.2. The irregularity is important since it enables difficult regions to be treated by a fine mesh while slowly varying regions only require a coarse mesh.

Consider one particular case of solving the Laplace equation

$$u_{xx} + u_{yy} = 0 \tag{8.31}$$

in a region R, with u given on the boundary $\mathscr{C}$ of R. It has been established that this solution is equivalent to minimizing the functional

$$\int_R (u_x{}^2 + u_y{}^2)\, dx\, dy. \tag{8.32}$$

It has been assumed that the region R has been triangulated as in figure 8.2 and attention will be concentrated on one particular triangle with vertices (x_0, y_0), (x_1, y_1), (x_2, y_2). The solution u of (8.31) is assumed in each triangle to be approximated by a linear function $u = ax + by + c$. To calculate this function for the particular triangle, first introduce the linear function $\alpha_0(x, y)$

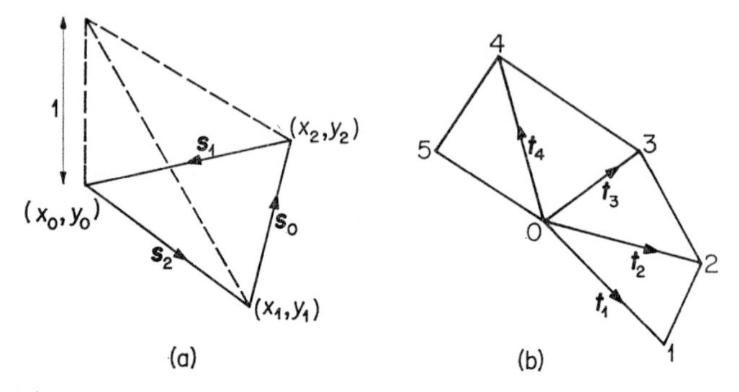

Figure 8.3 (a) Pyramid function, (b) Re-labelling of vectors.

which takes the values $\alpha_0(x_0, y_0) = 1$, $\alpha_0(x_1, y_1) = \alpha_0(x_2, y_2) = 0$. This is the pyramid function drawn in figure 8.3(a). The functions can be written down explicitly as

$$\alpha_0(x, y) = \begin{vmatrix} x & y & 1 \\ x_1 & y_1 & 1 \\ x_2 & y_2 & 1 \end{vmatrix} \div \begin{vmatrix} x_0 & y_0 & 1 \\ x_1 & y_1 & 1 \\ x_2 & y_2 & 1 \end{vmatrix} = \frac{1}{2A} \begin{vmatrix} x & y & 1 \\ x_1 & y_1 & 1 \\ x_2 & y_2 & 1 \end{vmatrix},$$

where the denominator is just twice the area (A) of the triangle.

Defining $\alpha_1(x, y)$ and $\alpha_2(x, y)$ similarly, the required linear function, taking the values u_0, u_1, u_2 at the respective corners, is just

$$u = u_0\alpha_0(x, y) + u_1\alpha_1(x, y) + u_2\alpha_2(x, y).$$

The u_i are unknown and must be chosen to satisfy the Laplace equation (at least approximately) or equivalently to minimize (8.32).

As in figure 8.3(a) the vectors along the sides of the triangle are

$$\mathbf{s}_0 = (x_2 - x_1, y_2 - y_1), \qquad \mathbf{s}_1 = (x_0 - x_2, y_0 - y_2),$$
$$\mathbf{s}_2 = (x_1 - x_0, y_1 - y_0)$$

and three new vectors are defined as

$$\mathbf{s'}_0 = (y_1 - y_2, x_2 - x_1), \qquad \mathbf{s'}_1 = (y_2 - y_0, x_0 - x_2),$$
$$\mathbf{s'}_2 = (y_0 - y_1, x_1 - x_0).$$

These three vectors are perpendicular to their respective $\mathbf{s}_i$ and the important property for the present analysis is that

$$\mathbf{s}_i \cdot \mathbf{s}_j = \mathbf{s}'_i \cdot \mathbf{s}'_j. \tag{8.33}$$

These vectors are required since

$$\nabla\alpha_0 = \frac{(y_1 - y_2, x_2 - x_1)}{2A} = \frac{\mathbf{s}'_0}{2A}$$

and hence in the triangle

$$\nabla u = \frac{u_0\mathbf{s}'_0 + u_1\mathbf{s}'_1 + u_2\mathbf{s}'_2}{2A}.$$

Now evaluate the integral (8.32) over the particular triangle being considered

$$I_\Delta = \int_\Delta \frac{1}{4A^2}(u_0\mathbf{s}'_0 + u_1\mathbf{s}'_1 + u_2\mathbf{s}'_2)^2\,dx\,dy$$

$$= \frac{(u_0\mathbf{s}'_0 + u_1\mathbf{s}'_1 + u_2\mathbf{s}'_2)^2}{4A}.$$

The integration is immediate since the integrand has no dependence on x or y. The total integral (8.32) will take the form $I = \Sigma I_\Delta$, summed over all triangles. Since the minimization will involve $\partial I/\partial u_k$ evaluate

$$\frac{\partial I_\Delta}{\partial u_0} = \frac{1}{2A}\mathbf{s}'_0 \cdot (u_0\mathbf{s}'_0 + u_1\mathbf{s}'_1 + u_2\mathbf{s}'_2)$$

$$= u_0\left(\frac{\mathbf{s}_0 \cdot \mathbf{s}_0}{2A}\right) + u_1\left(\frac{\mathbf{s}_0 \cdot \mathbf{s}_1}{2A}\right) + u_2\left(\frac{\mathbf{s}_0 \cdot \mathbf{s}_2}{2A}\right),$$

where (8.33) have been used.

Having established this key result it now becomes convenient to change the notation to that of figure 8.3(b) so that

$$\mathbf{s}_0 = \mathbf{t}_2 - \mathbf{t}_1, \qquad \mathbf{s}_1 = -\mathbf{t}_2, \qquad \mathbf{s}_2 = \mathbf{t}_1$$

and to distinguish this particular triangle the area is written A_{12}. The above equation now takes the form

$$\frac{\partial I_\Delta}{\partial u_0} = (u_1 - u_0)\left(\frac{\mathbf{t}_2 \cdot (\mathbf{t}_1 - \mathbf{t}_2)}{2A_{12}}\right) + (u_2 - u_0)\left(\frac{\mathbf{t}_1 \cdot (\mathbf{t}_2 - \mathbf{t}_1)}{2A_{12}}\right).$$

Since u_0 occurs only in the N triangles containing the point 0, for the minimum of $I = \Sigma I_\Delta$

$$\frac{\partial I}{\partial u_0} = \sum_{i=1}^{N} \omega_i(u_i - u_0) = 0, \tag{8.34}$$

where cyclic notation is used so that $\mathbf{t}_{N+1} \equiv \mathbf{t}_1$ etc, and

$$\omega_i = \tfrac{1}{2}\left(\frac{\mathbf{t}_{i+1} \cdot (\mathbf{t}_i - \mathbf{t}_{i+1})}{A_{i,i+1}} + \frac{\mathbf{t}_{i-1} \cdot (\mathbf{t}_i - \mathbf{t}_{i-1})}{A_{i-1,i}}\right). \tag{8.35}$$

It will be noted that the ω_i depend only on the geometry of the neighbouring triangles and each of the terms in (8.35) can be calculated as the cotangent of the angles in the triangles opposite the vector $\mathbf{t}_i$ (see problem 13).

The method is now complete: firstly choose the triangulation, secondly compute the ω_i from (8.35), thirdly write down the linear equation (8.34) for each of n mesh points giving n equations in the n unknowns u_i and finally solve these equations by any of the standard methods.

Example 60 Solve the Laplace equation in the region shown in figure 8.4 with $u = 1$ on the segments AB and DE inclusive of the end points and $u = 0$ on the rest of the boundary.

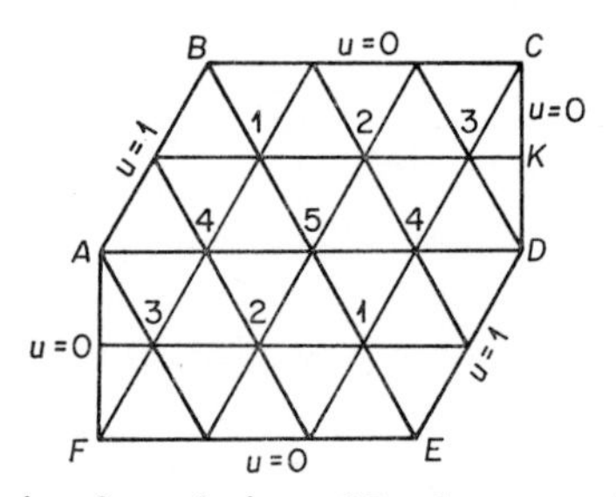

Figure 8.4 Region for solution of Laplace equation, example 60.

A simple mesh is drawn on the figure to illustrate the use of the method and symmetry has been used to reduce the number of unknowns. For all the points except point 3, $\omega_i = -1.155$; for point 3, $\omega_{3C} = \omega_{3D} = -0.577$, $\omega_{3K} = -3.464$ while $\omega_i = -1.155$ for the other three cases. Taking each of the points in turn the equations (8.34) become

$$0 = -1.155[(1 - u_1) + (1 - u_1) + (0 - u_1) + (u_2 - u_1) + (u_5 - u_1) + (u_4 - u_1)]$$

$$0 = -1.155[(u_1 - u_2) + (0 - u_2) + (0 - u_2) + (u_3 - u_2) + (u_4 - u_2) + (u_5 - u_2)]$$

$$0 = -1.155[(u_2 - u_3) + (0 - u_3) + \tfrac{1}{2}(0 - u_3) + 3(0 - u_3) + \tfrac{1}{2}(1 - u_3) + (u_4 - u_3)]$$

$$0 = -1.155[(1 - u_4) + (1 - u_4) + (u_1 - u_4) + (u_5 - u_4) + (u_2 - u_4) + (u_3 - u_4)]$$

$$0 = -1.155[(u_4 - u_5) + (u_1 - u_5) + (u_2 - u_5) + (u_4 - u_5) + (u_1 - u_5) + (u_2 - u_5)].$$

These equations can be solved very easily; taking reasonable first guesses at the solution a Gauss–Seidel iteration takes about four steps to give two figure accuracy

$$u_1 = 0.57, \quad u_2 = 0.31, \quad u_3 = 0.20, \quad u_4 = 0.60, \quad u_5 = 0.49.$$

It is clear that here only the simplest possible case has been studied. The present method can be generalized, however, in several ways: (1) a more general differential equation can be considered, (2) different boundary conditions need looking at, (3) the approximating functions can be made higher order polynomials, (4) the basic shape is normally a triangle but other basic shapes can be treated similarly. Interesting discussions of these points will be found, together with useful references, in Walsh (1971) or Zienkiewicz and Cheung (1967).

8.6* Non-linear Problems

In examples 33 and 34 one major complication of non-linear problems was noted. This was that even if an appropriate functional can be found the optimization leads to a set of non-linear algebraic equations to solve. These require computationally-sophisticated and usually slow hill climbing techniques to solve them. To illustrate this once more consider the Poisson–Boltzmann equation

$$u'' = \sinh u \tag{8.36}$$

with

$$u(0) = u_0, \qquad u(x) \to 0 \quad \text{as } x \to \infty.$$

The linearized version of this equation is obtained in the Debye–Hückel theory of electrolyte solutions and (8.36) can be considered as a more realistic generalization. It can be checked that the functional

$$I[u] = \int_0^\infty (\tfrac{1}{2}u'^2 + \cosh u - 1)\, dx$$

has the correct Euler equation; the (-1) term is introduced since as $u \to 0$ as $x \to \infty$, $\cosh u \to 1$ and the integral would be unbounded without it. The linearized version of (8.36) obviously has a solution $u = u_0 e^{-x}$ so it is natural to try as a first guess at the solution of (8.36) which satisfies the boundary conditions,

$$u = u_0 e^{-\alpha x}, \qquad \alpha > 0.$$

With this

$$I[u] = \frac{1}{4} u_0{}^2\alpha + \frac{1}{\alpha} K(u_0),$$

where

$$K(u_0) = \int_0^{u_0} \frac{(\cosh y - 1)}{y} \, dy,$$

which is independent of α and is a number depending only on u_0. The extremum with respect to α is now

$$\alpha = -\frac{2(K)^{\frac{1}{2}}}{u_0}.$$

This evaluation required the solution of a quadratic equation (albeit trivial in this case) but with any more adjustable constants the labour of the method mounts enormously.

Because of the large amount of work involved in higher approximations for the above method, other approaches are usually sought. Further comment will be made on this point in chapter 12, where some ideas suitable to the solution of non-linear problems are presented. Recently, however, solutions of non-linear differential equations have been found by iteratively solving a sequence of linear equations by fixed point or quasilinearization methods. These linear equations are usually solved numerically but a variational approximate technique is a possibility if explicit formulae are required.

To illustrate these methods consider (8.36) again. It is first assumed that an approximate solution $v(x)$ is known, then $u = v(x) + e(x)$ is tried as a solution with all squares and higher powers of e neglected. The equation (8.36) becomes therefore

$$v'' + e'' = \sinh v \cosh e + \cosh v \sinh e,$$

so that to first order in e

$$e'' = (\cosh v)e + (\sinh v - v''). \tag{8.37}$$

Since v is known, a linear equation in e has been obtained. The solution of (8.37) can still be a bit awkward so often this equation is simplified even further as

$$e'' = e + (\sinh v - v'').$$

This provides a so-called *fixed point* method which is usually described iteratively as

$$e''_{n+1} = e_{n+1} + (\sinh u_n - u''_n), \tag{8.38}$$

with

$$u_{n+1} = u_n + e_{n+1}$$

and u_n known from a previous calculation. Provided a first guess, $u_0(x)$ satisfying the boundary conditions, is supplied, then e_1 can be computed from (8.38) and hence u_1, the process continuing iteratively in this way. The boundary conditions imposed on $e_{n+1}(x)$ are that $e_{n+1}(0) = 0$ and $e_{n+1}(x) \to 0$ as $x \to \infty$ so that the new u_{n+1} still satisfies the correct conditions. It can be observed that provided $e_n \to 0$ as $n \to \infty$ equation (8.38) gives that u_n satisfies progressively more accurately the required equation (8.36).

The equivalent *quasilinearization* technique uses (8.37) directly as

$$e''_{n+1} = (\cosh u_n)e_{n+1} + (\sinh u_n - u''_n) \tag{8.39}$$

and the procedure is just as described above.

As an example of the use of (8.38) try as a first guess $u_0(x) = u_0 e^{-x}$ then the correct boundary conditions are satisfied and the solution is required of

$$e''_1 - e_1 = \sinh(u_0 e^{-x}) - u_0 e^{-x}$$

with

$$e_1(0) = 0, \qquad e_1(x) \to 0 \quad \text{as } x \to \infty.$$

This equation must be integrated numerically (or possibly analytically in this first case), then $u_1 = u_0 e^{-x} + e_1(x)$ and the iteration proceeds.

PROBLEMS

1. Find the adjoint of the operators

(i) $L \equiv x\dfrac{d^2}{dx^2} - \dfrac{d}{dx} + x,$ (ii) $M \equiv x\dfrac{\partial^2}{\partial x^2} + y\dfrac{\partial^2}{\partial y^2}$

(iii) $N \equiv \nabla^2 + k^2.$

2. Find the function $g(x)$ for which

$$g(au'' + bu' + cu) \equiv (Au')' + Cu,$$

thus making the operator self-adjoint.

3. Use the technique of example 54 to derive a suitable functional that has the biharmonic equation

$$(\nabla^2)^2 u = 0$$

as its Euler equation.

4. Estimate the lowest eigenvalue of the Bessel equation of order one

$$0 = x^2u'' + xu' + (\lambda x^2 - 1)u,$$

by using the functional

$$\int_0^1 \left(xu'^2 + \frac{u^2}{x}\right) dx \quad \text{subject to} \int_0^1 xu^2 \, dx = 1$$

and $u(0) = u(1) = 0$. Use the Rayleigh–Ritz procedure with $u = Ax(1 - x)$.

5. Show that the equation

$$\frac{\partial^2 u}{\partial t^2} = 4u \cos^2 x + \frac{\partial^2 u}{\partial x^2},$$

with $u = 0$ at $x = \pm\frac{1}{2}\pi$, has unstable modes. Use an analysis similar to section 8.1.4, with the function $u = e^{i\omega t} \cos x$, to prove this.

6. Use the Rayleigh–Ritz procedure, as described in section 8.2.1, to find the first two eigenvalues of

(i) $u'' + \lambda u = 0, \qquad |x| \leq \pi, \qquad x(-\pi) = x(\pi) = 0,$

(ii) $(2 - x^2)u'' - u + \lambda u = 0, \qquad |x| \leq 1, \qquad u(1) = u(-1) = 0.$

7. Given that L, M are self-adjoint operators, show that an estimate of the eigenvalues of

$$L\psi = \lambda M\psi \quad \text{in } R,$$

with $\psi = 0$ on the boundary of R, is obtained by minimizing the quotient

$$J[\varphi, \psi] = \frac{\int_R \varphi L\psi \, dV}{\int_R \varphi M\psi \, dV}.$$

8. Show that to satisfy the Poisson equation

$$\nabla^2\varphi = f \quad \text{in } R$$

and $\partial\varphi/\partial n = g$ on S, the boundary of R, the condition

$$\int_R f \, dV = \int_S g \, dS$$

must hold.

9. Show that solving the equation

$$xu'' + u' + xu = 0$$

with boundary conditions $u(0) = 0$, $u'(1) = 1$ is equivalent to finding the extremum of

$$\int_0^1 x(u'^2 - u^2) \, dx - 2u(1).$$

10. Modify the basic functional (8.25) to solve the Laplace equation $\nabla^2 u = 0$ in the square $|x| \leq a$, $|y| \leq a$, with $\partial u/\partial x = u$ on $y = a$ and $u = 1$ on the other three sides. Use this, with a suitable trial function, to find an approximate solution.

11. Suppose

$$\frac{\partial^2 u}{\partial x^2} = \frac{\partial u}{\partial t}$$

in $\{R: 0 < x < l,\ 0 < t < T\}$; show that the maximum (and similarly the minimum) of u on $\{\bar{R}: 0 \leq x \leq l, 0 \leq t \leq T\}$ occurs on one of the sides of $\bar{R}$, $x = 0$ or $x = l$ or $t = 0$.

12. If $\nabla^2 \varphi = 0$ in a closed region τ and $\varphi = f(P)$ for $P \in S$ (the surface of τ), show that

$$\int_\tau (\nabla\varphi)^2\, dV \leq \int_\tau (\nabla\psi)^2\, dV,$$

where ψ is any function with continuous second derivatives for which $\psi = f(P)$ for $P \in S$.

13. Show that the $(-\omega_i)$ of (8.35) is equal to the sum of the cotangents of the two angles opposite the $\mathbf{t}_i$ vector, in the triangles containing $\mathbf{t}_i$.

14. Use the finite element method to solve the Laplace equation in the region illustrated in figure 8.5 with the boundary conditions and mesh given.

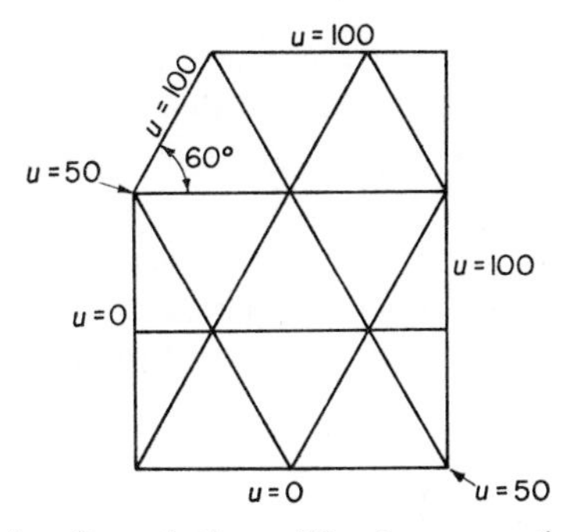

Figure 8.5 Region for solution of Laplace equation, problem 14.

Chapter IX*

Physical Applications

The present chapter involves the solution of some physical problems. It demands therefore that the reader has at least some elementary knowledge and experience of the application in hand. A lack of such knowledge would make the chapter a difficult one, so for readers in this position the final equations will have to be assumed without the background physics.

9.1* Fluid Mechanics

There have been many successful attempts to solve problems in fluid mechanics using a variational technique. These are well illustrated at an advanced level in the book by Schechter (1967) where the interest is mainly concentrated on research problems. Here two much more straightforward problems are treated to illustrate the use of the techniques. First it is necessary to quote the equations of fluid mechanics in their standard form. The first equation defines the conservation of mass in terms of the density ρ and velocity $\mathbf{v}$.

Continuity equation

$$\frac{\partial \rho}{\partial t} + \operatorname{div}(\rho \mathbf{v}) = 0. \tag{9.1}$$

The second equation is derived from Newton's second law, giving the external force/unit mass $\mathbf{F}$, the pressure p, and the viscous forces (viscosity μ) in terms of the acceleration

Momentum equation

$$\frac{\partial \mathbf{v}}{\partial t} + \mathbf{v} \,.\, \nabla \mathbf{v} = -\frac{1}{\rho} \nabla p - \frac{\mu}{\rho} \operatorname{curl} \operatorname{curl} \mathbf{v} + \mathbf{F}. \tag{9.2}$$

If the flow is compressible, that is ρ is not constant, then these equations must be supplemented by an *equation of state* which relates p and ρ; this depends on the fluid under consideration and is obtained from thermodynamic experiments. The equations must of course be solved together with suitable boundary conditions. The most important of these is the condition at a solid boundary, firstly there must be no flow through the boundary and secondly for a viscous fluid ($\mu \neq 0$) the fluid has zero tangential velocity at this boundary.

9.1.1* FLOW DOWN A PIPE

It is assumed in this problem that a viscous incompressible fluid flows steadily down a straight pipe of arbitrary cross-section under a pressure gradient. These assumptions imply $\mu \neq 0$, $\rho =$ constant and $\partial/\partial t \equiv 0$. The configuration under consideration is illustrated in figure 9.1. Using the coordinate system illustrated in the figure a further assumption that the flow is

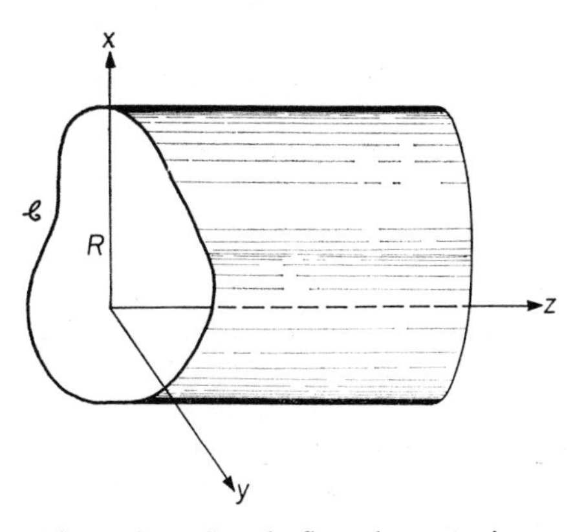

Figure 9.1 Steady flow down a pipe.

parallel to the z-axis is also made; this implies that $\mathbf{v} = \{0, 0, v(x, y)\}$. The equation (9.1) is now immediately satisfied, while the x and y-components of the momentum equation (9.2) are satisfied if and only if $p = p(z)$. The final z-component of (9.2) now takes the form

$$0 = \rho v \frac{\partial v(x, y)}{\partial z} = -\frac{dp}{dz} - \mu \nabla^2 v.$$

The left-hand side of this equation is put to zero since v is independent of z and since $\nabla^2 v$ is also independent of z then $dp/dz = -P$ must be a constant.

Hence

$$\frac{\partial^2 v}{\partial x^2} + \frac{\partial^2 v}{\partial y^2} = -\frac{P}{\mu}. \tag{9.3}$$

The basic problem therefore is to solve this Poisson equation for v in the region R bounded by the curve $\mathscr{C}$. On $\mathscr{C}$, $v = 0$ since a viscous fluid is being considered. For simple geometries this equation can be solved explicitly but failing this a numerical solution is straightforward. An alternative method is to write (9.3) as an optimization problem. The Euler equation of the functional

$$J[v] = \iint_R \left(v_x{}^2 + v_y{}^2 - \frac{2Pv}{\mu} \right) dx\, dy \tag{9.4}$$

was shown in section 8.3, using equation (8.25), to be identical with (9.3).

Take as an example the case of flow through a square duct

$$R: |x| \leq a, \qquad |y| \leq a.$$

A simple approximation that satisfies the boundary condition $v = 0$ on $\mathscr{C}$ is

$$v = A(a^2 - x^2)(a^2 - y^2).$$

Substituting into (9.4) and performing the integrations gives

$$J[v] = \frac{256}{45} A^2 a^8 - \frac{32}{9} \frac{Pa^6}{\mu} A,$$

and the extremum with respect to A is calculated as $A = 5P/16(\mu a^2)$. As a check of the accuracy at $x = y = 0$ a numerical solution shows $v(0, 0) = 0.295(Pa^2/\mu)$ while the present approximation produces the result $v(0, 0) = 0.312(Pa^2/\mu)$, which is within 5%. Frequently the main interest of such problems is not the detailed flow profiles but an average property which can often be calculated accurately using a comparatively crude approximation. For instance, if the throughflow is required in this problem, the mass flowing through a section of the duct in unit time is

$$M = \rho \iint_R v\, dx\, dy.$$

The numerically obtained value to M is 0.560 $(\rho Pa^4/\mu)$ while this variational approximation yields $M = 0.556(\rho Pa^4/\mu)$. This compares very favourably giving an accuracy of less than 1%.

9.1.2* COMPRESSIBLE FLOW

An illustration of a variational technique applied to a more difficult compressible flow problem is provided by a paper by Lush and Cherry (1956).

The flow is assumed to be steady, $\partial/\partial t \equiv 0$, inviscid, $\mu = 0$, with zero external forces and irrotational. This last assumption implies curl $\mathbf{v} = 0$ (or that there is no rotation) which by a standard theorem in vector analysis (see Bourne and Kendall (1967)) implies that $\mathbf{v} = \nabla\varphi$. The continuity equation (9.1) just becomes

$$\text{div}\,(\rho\mathbf{v}) = 0, \tag{9.5}$$

while the momentum equation (9.2) with $\mu = 0$ can be integrated to give the Bernouilli equation

$$\tfrac{1}{2}\mathbf{v}^2 + \int \frac{dp}{\rho} = \text{constant} \tag{9.6}$$

and the equation of state necessary for a compressible flow is taken to be

$$p = p(\rho). \tag{9.7}$$

Given φ, $\mathbf{v} = \nabla\varphi$ can be calculated, ρ is then computed from (9.6) and finally p from (9.7). The continuity equation (9.5) will not of course be satisfied in general and Lush and Cherry attempt to construct a functional which, when optimized, will ensure the satisfaction of (9.5). They consider the functional

$$J[\varphi] = \int_\tau p\, dV, \tag{9.8}$$

where τ is region under consideration with boundary S. Because of the complexity of the relation between p and φ it is not possible to treat this functional by standard methods and it is necessary to resort to first principles, albeit following the general ideas that have already been developed. Leaving boundary conditions aside for the moment suppose that $\bar{\varphi}$ gives the extremum of (9.8) (*note* Corresponding extreme values are denoted by $\bar{\mathbf{v}}$, $\bar{\rho}$, etc) and consider variations

$$\varphi = \bar{\varphi} + \varepsilon\psi \tag{9.9}$$

with corresponding changes $\rho = \bar{\rho} + \varepsilon\sigma$ and $p = \bar{p} + \varepsilon P$. To first order in ε (9.7) gives $P = p'(\bar{\rho})\sigma$. The term $U = \int dp/\rho$ in (9.6) gives

$$U(\bar{\rho} + \varepsilon\sigma) - U(\bar{\rho}) = \int_{\bar{\rho}}^{\bar{\rho}+\varepsilon\sigma} \frac{p'(z)}{z}\, dz$$

$$= \frac{p'(\bar{\rho})\varepsilon\sigma}{\bar{\rho}} = \frac{P\varepsilon}{\bar{\rho}}.$$

The equation (9.6) now becomes

$$\tfrac{1}{2}\bar{\mathbf{v}}^2 + \varepsilon\bar{\mathbf{v}} \,.\, \nabla\psi + \int \frac{d\bar{p}}{\bar{\rho}} + \varepsilon\frac{P}{\bar{\rho}} = C$$

or

$$P = -\bar{\rho}\bar{\mathbf{v}} \,.\, \nabla\psi.$$

Using this result in (9.8) to first order in ε the equation reduces to

$$F(\varepsilon) = J[\varphi] = J[\bar{\varphi}] - \varepsilon \int_\tau \bar{\rho}\bar{\mathbf{v}} \,.\, \nabla\psi \, dV.$$

Since it may be assumed (as usual) that ψ, $\bar{\rho}$, $\bar{\mathbf{v}}$, etc are known functions, $J[\varphi]$ is just a function of ε, $F(\varepsilon)$, with $\varepsilon = 0$ as an extremum and hence $F'(0) = 0$. Thus

$$\begin{aligned} 0 &= \int_\tau \bar{\rho}(\bar{\mathbf{v}} \,.\, \nabla\psi) \, dV \\ &= \int_\tau \{\mathrm{div}\,(\bar{\rho}\psi\bar{\mathbf{v}}) - \psi \,\mathrm{div}\,(\bar{\rho}\bar{\mathbf{v}})\} \, dV \\ &= \int_S \psi\bar{\rho}\bar{\mathbf{v}} \,.\, \mathbf{dS} - \int_\tau \psi \,\mathrm{div}\,(\bar{\rho}\bar{\mathbf{v}}) \, dV. \end{aligned}$$

If the velocities on the boundary S of τ are such that $\bar{\mathbf{v}} \,.\, \hat{\mathbf{n}} = 0$ then the first integral is zero and since ψ was chosen arbitrarily the Euler equation just becomes

$$\mathrm{div}\,(\bar{\rho}\bar{\mathbf{v}}) = 0.$$

Therefore, if $\bar{\mathbf{v}} \,.\, \hat{\mathbf{n}} = 0$ on S, optimizing (9.8) over all φ using the construction described ensures that the continuity equation (9.5) is satisfied.

The boundary conditions just obtained are very restrictive and they can be relaxed by considering a modified functional

$$J_1[\varphi] = \int_\tau p \, dV + \int_S \rho\varphi\mathbf{v} \,.\, \mathbf{dS}. \tag{9.10}$$

The Euler equation of this functional still remains the same (i.e. the continuity equation) but the boundary conditions become $(\rho\mathbf{v} \,.\, \hat{\mathbf{n}})$ prescribed on the boundary. The reader is referred to the original paper for the details of this deduction and also for the detailed application to a realistic problem.

It may be reasonably asked how the functionals (9.8) or (9.10) were constructed in the first place. Once they have been written down the deduction of the Euler equation is quite straightforward. The whole process can be compared with the process of differentiation and integration. The derivation of the Euler equation from a given functional is essentially a functional differentiation which can always be performed provided sufficient continuity conditions are satisfied. On the other hand obtaining the functional from the Euler equation can be thought of as functional integration; it should be recalled that some integrals of standard form can be integrated easily, some can be manipulated to standard form while some cannot be integrated at all in terms of simple functions. Likewise certain classes of differential equations can be derived from known functionals (cf. chapter 8), some can be done from first principles (as in the above problem) while for most of them it is quite impossible to find the corresponding functional.

9.2* Elasticity, Thin Beams

The technological theory of thin beams considers small transverse displacements of a uniform beam whose cross-sectional dimensions are negligible compared with the length dimension. Suppose that the beam is set up with coordinates indicated in figure 9.2(a) and loaded by a force $f(x)$/unit length.

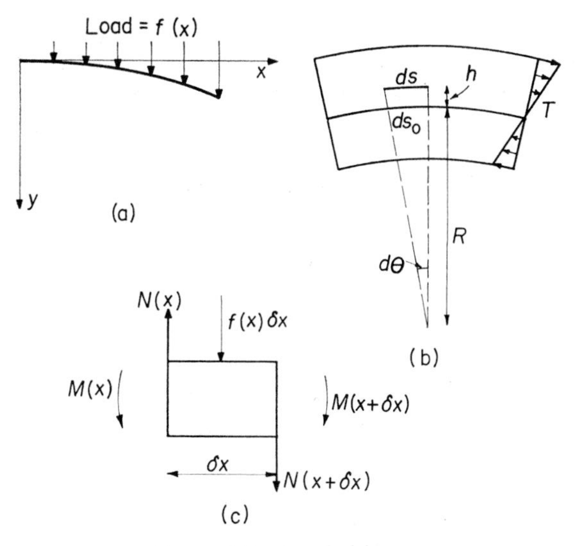

Figure 9.2 Loaded beam.

Looking at a small element of the beam, figure 9.2(b), the extension ds, relative to the centre line which is assumed to remain unstretched, is given by

$$\frac{ds - ds_0}{ds_0} = \frac{(R + h)\delta\theta - R\delta\theta}{R\delta\theta} = \frac{h}{R},$$

where R is the radius of curvature of the beam at that element. The Hookean approximation that force is proportional to extension is used to derive the tensile force T in the cross-section as

$$T = \frac{Eh}{R},$$

where E is Young's modulus. Now this force, which varies linearly with h through the section as indicated in figure 9.2(b), gives a bending moment M about the centre line. This moment can be calculated as

$$M = \iint Th\, dS = \frac{E}{R}\iint h^2\, dS = \frac{EI}{R},$$

where I is the moment of inertia of the cross-section about an axis through the centre line perpendicular to the section. The geometrical assumption that the flexure is small is used to neglect $(dy/dx)^2$ compared to unity, giving

$$R = \frac{(1 + y'^2)^{3/2}}{y''} \approx \frac{1}{y''}.$$

The equilibrium of the element can now be considered from figure 9.2(c). In addition to the bending moment there is of course a shear force N. Equating forces in the figure

$$N(x + \delta x) - N(x) + f(x)\delta x = 0$$

or

$$\frac{dN}{dx} = -f(x).$$

Taking moments about the centre

$$M(x + \delta x) - M(x) + N(x)\delta x = 0$$

or

$$\frac{dM}{dx} = -N.$$

Combining these two equations produces the result

$$\frac{d^2M}{dx^2} = f(x)$$

and using $M = EIy''$ gives a final equation for the displacement y as

$$EIy'''' = f(x). \tag{9.11}$$

The functional corresponding to this beam equation is

$$J[y] = \int_0^l [\tfrac{1}{2}EIy''^2 - yf(x)]\,dx,$$

as shown in example 43. The two terms in this functional can be identified as the strain energy density and the work done by the external forces. The strain energy measures the energy stored in the beam because of the deformation and is computed from $\frac{1}{2}$(stress) $\times$ (strain) $= \frac{1}{2}(Eh/R) \times (h/R)$. (*Note* In this approximation the energy due to the shear forces is assumed to be negligible.) When this energy is summed over the whole cross-section and $R \approx 1/y''$ is used the total strain energy density for the cross-section becomes $\frac{1}{2}EIy''^2$. This functional was used in examples 27 and 43 to solve for the equilibrium configurations of the beam.

9.3* Relativity

Again it would be out of place in this book to study all the detailed physics required for a full understanding of relativity. Just sufficient will be quoted to give the flavour of the arguments and the mathematics used.

Referred to one person's inertial frame and clock the totality of all events is given by the coordinates $x_1, x_2, x_3, x_4 = t$. For adjacent events the interval, ds, between the two events is measured by the metric

$$ds^2 = \sum_{ij} g_{ij}\,dx_i\,dx_j$$

with $g_{ij} = g_{ji}$ (without loss of generality) and all being differentiable functions of the x_i's. In *Newtonian mechanics*

$$ds^2 = dx^2 + dy^2 + dz^2 \tag{9.12}$$

while in *special relativity*

$$ds^2 = dx^2 + dy^2 + dz^2 - c^2\,dt^2, \tag{9.13}$$

where c is the velocity of light. In *general relativity* a wider class of metrics must be considered; they must, of course, satisfy several requirements but these will not be quoted here. One particularly interesting case will be studied, this is the *Schwartzschild metric*

$$ds^2 = \frac{dr^2}{[1 - (2m/r)]} + r^2(d\theta^2 + \sin^2\theta\,d\varphi^2) - c^2\left(1 - \frac{2m}{r}\right)dt^2. \tag{9.14}$$

This corresponds to a massive, spherically symmetric, static, gravitating body in free space, that is the metric relates to an isolated star.

Newton's first law of motion states that a body under no forces continues in a state of rest or uniform motion. This is replaced in relativity by the law that particles travel along geodesics in the space. Since in relativity light travels with a finite velocity the light path must also be computed. From (9.13) it may be noted that

$$\dot{s}^2 = V^2 - c^2$$

and hence if $V = c$ (i.e. the test particle is travelling with the velocity of light) then $ds = 0$. Thus in special relativity light travels along *null* geodesics and this is generalized into a similar hypothesis in general relativity.

Consider the path of a small test particle (e.g. a planet) in the neighbourhood of a star given by the metric (9.14). To find this orbit it is necessary to construct the geodesic equations obtained in section 7.4 as

$$x''_p + \sum_{j,k} \left\{ \begin{matrix} p \\ j \quad k \end{matrix} \right\} x_j x_k = 0$$

and then to solve these equations. The Christoffel symbols can be constructed and the θ, φ and t equations can be computed

$$\theta'' + \frac{2}{r} r'\theta' - \sin\theta\cos\theta\,\varphi'^2 = 0 \tag{9.15}$$

$$\varphi'' + \frac{2}{r} r'\varphi' - 2\cot\theta\,\theta'\varphi' = 0 \tag{9.16}$$

$$t'' + \frac{2m}{r(r - 2m)} r't' = 0 \tag{9.17}$$

where dash denotes differentiation with respect to the arc length s. The fourth equation for the r-coordinate is complicated and may be replaced by the metric itself

$$1 = \frac{r'^2}{[1 - (2m/r)]} + r^2(\theta'^2 + \sin^2\theta\,\varphi'^2) - c^2\left(1 - \frac{2m}{r}\right) t'^2. \tag{9.18}$$

To solve these four equations first note that (9.16) can be written

$$\frac{d}{ds}(r^2 \sin^2\theta\,\varphi') = 0$$

and assuming (without loss of generality) that $\varphi = \varphi' = 0$ at $t = 0$ then $\varphi \equiv 0$ for all time. Equation (9.15) now simplifies and integrates to

$$r^2\theta' = \text{constant} = \alpha, \tag{9.19}$$

while (9.17) can be written

$$\frac{d}{ds}\left(\frac{(r - 2m)t'}{r}\right) = 0$$

so that

$$t' = \frac{\beta r}{r - 2m}, \tag{9.20}$$

where β is an arbitrary constant. The final equation (9.18) is best described using the coordinates $u = 1/r$ and replacing s by $d/ds \equiv \alpha u^2 \, d/d\theta$ from (9.19), whence

$$\alpha^2 \left(\frac{du}{d\theta}\right)^2 = (1 - \alpha^2 u^2)(1 - 2mu) - c^2\beta^2.$$

Differentiating with respect to θ

$$\frac{d^2u}{d\theta^2} + u = -\frac{m}{\alpha^2} + 3mu^2.$$

Since (9.19) is basically the conservation of angular momentum, in terms of real time $h = i\alpha c$; the real mass M is identified as $m = M\gamma/c^2$, where γ is the gravitational constant, and the final equation becomes

$$\frac{d^2u}{d\theta^2} + u = \frac{\gamma M}{h^2} + \frac{3M\gamma u^2}{c^2}.$$

The final term of this equation is the well known *Einstein correction term* to the Newtonian orbit of the planets. The equation may be solved by successive approximation, the first approximation obtained by neglecting the Einstein term gives the standard elliptic orbit $u = (\gamma M/h^2)(1 + e \cos \theta)$. Including the correction term, the second approximation, with only terms in e retained, can be calculated as

$$u = \frac{\gamma M}{h^2}[1 + e \cos(\theta - \delta\theta)],$$

where $\delta\theta = [3(M\gamma)^2/c^2h^2]\theta$. Although this correction is very small it implies that the axes of the elliptic orbit of a planet precess slowly. This precession has been measured for Mercury and is found to be about 43 seconds of arc per century, when corrections have been made for perturbations due to other planets. This result agrees well with the computed value and constitutes one of the major successes of general relativity.

The second major success was to predict that a light ray passing close to the sun would be deflected due to relativistic effects. This was observed at an eclipse in 1919 and the calculated value found to be reasonably accurate. The theoretical prediction is again obtained by solving (9.15), (9.16), (9.17) together with the equation corresponding to (9.18) for a null geodesic

$$0 = \frac{r'^2}{1 - 2mu} + r^2(\theta'^2 + \sin^2 \theta\, \varphi'^2) - c^2(1 - 2mu)t'^2.$$

Some care must be taken since it is not possible to use arc length s as the parameter since $s = 0$ along the whole of a null geodesic. The analysis follows almost identically, however, and gives the equation of the path as

$$\frac{d^2u}{d\theta^2} + u = \frac{3M\gamma u^2}{c^2}.$$

The first approximation to the solution is $u = \cos\theta/R$ or a straight line distance R from the centre of the sun, while the second approximation is $u = (\cos\theta/R) + (M\gamma/R^2c^2)(2 - \cos^2\theta)$ which determines the relativistic deflection. The details of this calculation are left to the reader.

9.4* Quantum Mechanics

The basic problem in quantum mechanics is to obtain a solution of the Schrodinger equation

$$H\psi = E\psi, \tag{9.21}$$

where H is a differential operator called the Hamiltonian and E is an eigenvalue or energy level. For a single particle of mass m in a potential energy field $V(x, y, z)$ the Hamiltonian takes the form

$$H \equiv -\frac{\hbar^2}{2m}\nabla^2 + V,$$

where $h(\hbar = h/2\pi)$ is Planck's constant. In general the Hamiltonian is obtained from the corresponding classical Hamiltonian $H = T + V$ (T = kinetic energy) by replacing the linear momentum $\mathbf{p}$ by $(\hbar/i)\nabla$, position vector $\mathbf{r}$ remains the same and other variables, such as angular momentum, transform suitably. The operator H is not necessarily real but it can be shown to be Hermitian, that is for all functions ψ (in general complex) which are suitably differentiable

$$\int (H\psi)^*\psi\, d\tau = \int \psi^* H\psi\, d\tau,$$

where * denotes complex conjugate and the integrations are over the whole of the available space.

Complete determinism is lost in quantum mechanics and a probabilistic approach must be used. For a single particle problem the basic assumption is that $(\psi^*\psi)\,dx\,dy\,dz$ is the probability that on measurement the particle will be found in the volume element $(dx\,dy\,dz)$. Thus $\psi^*\psi$ is a probability density which must, of course, satisfy the normalization condition

$$\int \psi^*\psi\,d\tau = 1.$$

The expectation value of any operator F (e.g. position, momentum, etc) is obtained from

$$\langle F\rangle = \int \psi^* F\psi\,d\tau.$$

In the case when $\psi = \psi_n$ is an eigenfunction of (9.21) so that

$$H\psi_n = E_n\psi_n$$

the expectation value of H is

$$\langle H\rangle = \int \psi^*{}_n H\psi_n\,d\tau = \int E_n\psi^*{}_n\psi_n\,d\tau = E_n,$$

or just the appropriate energy level. It is only when the system is in an *eigenstate* that the probability distribution of F is peaked at a definite value with no spread. In particular the value of H would be quite definite at E_n. These eigenvalues are normally discrete, certainly in the cases considered here, but they can of course be degenerate so that several independent eigenfunctions can give exactly the same eigenvalue. This causes some complications in the algebra but using the Gram–Schmidt process (see section 2.3.4) the eigenfunctions can be shown to form a complete orthonormal set satisfying

$$\int \psi^*{}_i\psi_j\,d\tau = \delta_{ij}. \tag{9.22}$$

If a system is not in an eigenstate then the wave function ψ can be expanded in terms of these complete eigenfunctions as $\psi = \Sigma\, c_i\psi_i$ and then the c_i interpreted suitably. Without going into detail over this non-trivial physical interpretation, suffice it to add that all the information about the system is included in the eigenvalues and eigenfunctions.

The Schrodinger equation (9.21), although linear, is a very complicated differential equation and only rarely can explicit solutions be obtained. Approximate techniques are therefore extremely important and direct variational methods of the Rayleigh–Ritz type play an important part in the theory. First an example which can be computed exactly.

Example 61 Find the eigenvalue and eigenfunction of the lowest energy state of the hydrogen atom with Coulomb potential $V(r) = -Ze^2/r$, where Z is the charge of the nucleus, e the electron charge and r the distance of the electron from the nucleus.

The Schrodinger equation (9.21) takes the form

$$\nabla^2\psi + \frac{2m}{\hbar^2}\left(E + \frac{Ze^2}{r}\right)\psi = 0. \tag{9.23}$$

While it is possible to solve this equation exactly in terms of Legendre and Laguerre polynomials, concentrate on the lowest energy state which can be proved to be independent of the angular variables so that $\psi = \psi(r)$. The equation (9.23) then becomes the ordinary differential equation

$$\frac{d^2\psi}{dr^2} + \frac{2}{r}\frac{d\psi}{dr} + \frac{2m}{\hbar^2}\left(E + \frac{Ze^2}{r}\right)\psi = 0. \tag{9.24}$$

Since the boundary conditions insist that ψ remains bounded for all finite r and $\psi \to 0$ as $r \to \infty$, the solutions of (9.24) take the form $\psi = (\text{polynomial}) \times \exp(-\alpha r)$, where α is a positive constant. The simplest of these functions occurs when the polynomial is a constant and this corresponds to the lowest energy state. Put

$$\psi = Ae^{-\alpha r}$$

into (9.24) then

$$Ae^{-\alpha r}\left(\alpha^2 - \frac{2\alpha}{r} + \frac{2mE_0}{\hbar^2} + \frac{2mZe^2}{\hbar^2}\frac{1}{r}\right) = 0$$

which is satisfied by $\alpha = mZe^2/\hbar^2$ and $E_0 = -\hbar^2\alpha^2/2m$. If the Bohr radius is defined by $a = \hbar^2/me^2$ these can now be rewritten

$$\alpha = \frac{Z}{a} \quad \text{and} \quad E_0 = -\frac{\frac{1}{2}Z^2me^4}{\hbar^2}.$$

Since ψ must be normalized to make $\int \psi^*\psi\, d\tau = 1$ this gives

$$1 = A^2\int_0^\infty e^{-2\alpha r}r^2 4\pi\, dr = \frac{\pi A^2}{\alpha^3}$$

and hence

$$\psi_0 = \left(\frac{Z^3}{\pi a^3}\right)^{\frac{1}{2}} \exp\left(-\frac{Zr}{a}\right), \qquad E_0 = -\frac{\frac{1}{2}Z^2me^4}{\hbar^2}.$$

For more complex problems such simple solutions are not available but the lowest eigenvalue can be computed by using the result

$$E_0 \leq \frac{\int \psi^* H \psi \, d\tau}{\int \psi^* \psi \, d\tau} \tag{9.25}$$

suggested in section 8.2.1. Computing the minimum of the RHS of (9.25) over a restricted set of ψ's gives an estimate of this lowest energy. If, of course, $\psi = \psi_0$ is the exact eigenfunction then equality is achieved in (9.25). To prove (9.25) expand ψ in terms of the eigenfunctions $\psi = \Sigma\, a_n \psi_n$ and use the orthonormality conditions (9.22)

$$\begin{aligned}\text{RHS} &= \frac{\Sigma\, a^*_m a_n \int \psi^*_m H \psi_n \, d\tau}{\Sigma\, a^*_m a_n \int \psi^*_m \psi_n \, d\tau} \\ &= \frac{\Sigma\, |a_n|^2 E_n}{\Sigma\, |a_n|^2}.\end{aligned}$$

Since $E_n \geq E_0$, the RHS $\geq E_0$ and hence (9.25) follows.

Example 62 Compute the ground state energy of the Helium atom.

Helium consists of a nucleus and two electrons with position vectors $\mathbf{r}_1$ and $\mathbf{r}_2$ relative to the nucleus. The Hamiltonian takes the form

$$H \equiv -\frac{1}{2}\frac{\hbar^2}{m}(\nabla_1{}^2 + \nabla_2{}^2) - Ze^2\left(\frac{1}{r_1} + \frac{1}{r_2}\right) + \frac{e^2}{r_{12}},$$

where the Laplacian operators refer to the 1 and 2 electrons and $r_{12} = |\mathbf{r}_1 - \mathbf{r}_2|$. An exact solution of the Schrodinger equation (9.21) with this Hamiltonian is quite out of the question and a variational method will be tried using (9.25).

The ground state wave function for the hydrogen atom was computed in example 61 as $A \exp(-Zr/a)$, so it appears reasonable that a product of two such expressions for the 1 and 2 electrons might suffice as an approximation to the required eigenfunction. So try

$$\psi = \frac{Z'^3}{\pi a^3} \exp\left(\frac{-Z'(r_1 + r_2)}{a}\right) \tag{9.26}$$

which has been normalized so that $\int \psi^2 \, d\tau = 1$ and Z' can now be adjusted to give the minimum value to the RHS of (9.25). The integral

$$\int \psi \left[-\frac{1}{2}\frac{\hbar^2}{m}(\nabla_1{}^2 + \nabla_2{}^2) - Ze^2\left(\frac{1}{r_1} + \frac{1}{r_2}\right)\right] \psi \, d\tau = \frac{e^2}{a}(Z'^2 - 2ZZ')$$

and is a straightforward exercise in integration. The remaining term

$\int \psi(e^2/r_{12})\psi\, d\tau$ is more difficult to evaluate but it can be checked that using the variables $u = r_1 + r_2$, $v = r_1 - r_2$, $w = r_{12}$ this integral gives $5Z'e^2/8a$. The total contribution is therefore

$$E_0 \leq \frac{e^2}{a}(Z'^2 - 2ZZ' + \tfrac{5}{8}Z').$$

The minimum of the quadratic in Z' is given by $Z' = Z - \frac{5}{16}$ so that using $Z = 2$ for Helium

$$E_0 \leq -\frac{e^2}{a}\left(\frac{27}{16}\right)^2 = 2E_H\left(\frac{27}{16}\right)^2,$$

where E_H is the lowest energy level of hydrogen computed above. Putting in values for the various constants gives a variational estimate of $E_0 = -77.46$ eV while the experimental value is -78.60 eV; this is an accuracy of less than 2% with a comparatively poor initial approximation.

An extension of this method can be used and it will be illustrated in a simple case (see section 8.2.1 for a similar general case). Suppose that two suitable approximate wave functions ψ_1 and ψ_2 are known then $\psi = c_1\psi_1 + c_2\psi_2$ is used in (9.25) to obtain a better approximation to the lowest energy level and gives

$$E(c_1, c_2) = \frac{\int (c_1\psi_1 + c_2\psi_2)^* H(c_1\psi_1 + c_2\psi_2)\, d\tau}{\int (c_1\psi_1 + c_2\psi_2)^*(c_1\psi_1 + c_2\psi_2)\, d\tau}.$$

Let

$$H_{ij} = \int \psi^*{}_i H\psi_j\, d\tau, \qquad S_{ij} = \int \psi^*{}_i\psi_j\, d\tau$$

then this equation can be rewritten

$$E(c_1, c_2)(S_{11}c_1{}^2 + 2S_{12}c_1c_2 + S_{22}c_2{}^2) = c_1{}^2H_{11} + 2c_1c_2H_{12} + c_2{}^2H_{22}.$$

The minimum of E with respect to c_1 and c_2 is obtained from

$$\frac{\partial E}{\partial c_1} = 0 \quad \text{gives } E(c_1S_{11} + c_2S_{12}) = c_1H_{11} + c_2H_{12}$$

$$\frac{\partial E}{\partial c_2} = 0 \quad \text{gives } E(c_1S_{12} + c_2S_{22}) = c_1H_{12} + c_2H_{22}.$$

Hence to get a non-trivial solution for c_1 and c_2

$$\begin{vmatrix} H_{11} - ES_{11} & H_{12} - ES_{12} \\ H_{12} - ES_{12} & H_{22} - ES_{22} \end{vmatrix} = 0.$$

This is normally called the *secular equation* and the lowest value of E satisfying this determinantal equation provides an estimate of the lowest energy level.

The extension of this procedure to include many trial wave functions is a very powerful tool particularly in the study of the energy levels of molecules and complex atoms. It is interesting to note that the higher roots of the secular equation provide estimates of the higher eigenvalues of the Schrodinger equation (see section 8.2.1 and Kemble (1958), p. 415).

PROBLEMS

1. In the square pipe flow problem of section 9.1.1, verify that an improved approximation is

$$v = (a^2 - x^2)(a^2 - y^2)[A + B(x^2 + y^2)],$$

where all the symmetries have been used. Evaluate A and B and compare $v(0, 0)$ and the mass flow with the values given in the text.
2. Try to set up an approximating sequence for the pipe flow problem when the cross-section is an equilateral triangle.
3. Verify that the functional (9.10) has natural boundary condition $(\rho\mathbf{v} \,.\, \hat{\mathbf{n}})$ prescribed on the boundary.
4. A cantilever beam is loaded by a force/unit length of $F_0 \sin(\pi x/l)$. At $x = 0$ the beam is clamped $y(0) = y'(0) = 0$, while at $x = l$ the beam is free so that $y''(l) = y'''(l) = 0$. Solve the beam equation (9.11) exactly and compare with a solution obtained using the Rayleigh–Ritz method on the corresponding functional.
5. Construct the geodesic equations and solve them for the special relativity metric (9.13).
6. Follow through the analysis leading to the equation (at the end of section 9.3) for a null geodesic in the Schwarzschild metric.
7. Repeat the relativistic calculation in section 9.3 on the de Sitter metric

$$ds^2 = \frac{dr^2}{[1 - (r^2/R^2)]} + r^2(d\theta^2 + \sin^2\theta\, d\varphi^2) - c^2\left(1 - \frac{r^2}{R^2}\right) dt^2.$$

8. The lowest energy of the Schrodinger equation for the hydrogen atom (9.24) was obtained in the text. Estimate the next energy level corresponding to wave functions dependent on r only.
9. The one dimensional harmonic oscillator has Schrodinger equation

$$-\frac{\hbar^2}{2\mu}\frac{d^2\psi}{dx^2} + (\tfrac{1}{2}\mu\omega^2 x^2 - E)\psi = 0.$$

Find the lowest energy level of this equation using a wave function of the form $\psi = \exp(-\beta x^2)$. (*Note* The equation has exact solutions in terms of

Hermite polynomials.) Estimate the effect on this level of including anharmonic terms $V(x) = \frac{1}{2}\mu\omega^2 x^2 + kx^3$.

10. Repeat an analysis similar to example 62 to obtain the lowest energy for the Yukawa potential

$$V = -V_0 \frac{\exp[-(r/a)]}{(r/a)}, \qquad V_0 > 0.$$

Try a solution of the form $\psi(r) = \exp(-\beta r/a)$, with β as a variable parameter.

Chapter X

Dynamic Programming

10.1 Discrete Problem

General variational problems have been studied already from two angles, the Euler equation and the Rayleigh–Ritz method. The first method produces exact solutions but only for a very limited number of physical problems, for example the brachistochrone of section 7.1. The second approach produced approximate solutions but this also ran into difficulties when the functionals are grossly non-linear, cf. example 34. In even more complicated problems, such as optimizing an earth–moon trajectory for a spacecraft, other severe difficulties occur because the governing differential equations of such problems cannot be solved explicitly. Since a numerical solution is inevitable, the question that may be asked is whether it would not be better to start the whole problem with a numerical approach. It was thoughts like these, together with the increasing power of digital computers to do the donkey work, that probably led to the development of the dynamic programming method in the 1950s under the leadership of R. E. Bellman. The subject has made tremendous strides over the past fifteen years and is well used in transportation, scheduling and space applications. All the examples quoted in this chapter are very simple, but most practical examples require a substantial report to set the problem up and a detailed computer implementation to solve them. It is hoped, therefore, that the examples quoted will give the essence of the method and the mathematics used.

The method may be described in its simplest form for a discrete multistage decision process. At each of a sequence of times $t_1, t_2, \ldots, t_N$, or *stages*, a decision has to be made to follow one of a finite number of paths to the next decision time, each of these paths having a given cost. The problem is to find the overall path from stage t_1 to stage t_N which minimizes the total cost. Recall the example quoted in chapter 1 where the path with minimum distance between London's King's Cross and Charing Cross stations was required. The decision points are the road junctions (which can be ordered suitably at sequential stages as required above), the decision to be taken is to turn left, right or to go straight on (or whatever the particular junction implies) and the cost of each decision is measured by the length of the road to the next junction. The overall policy is to select out of all possible routes the one with minimum length. In theory, provided the map is a finite one and the reasonable rule that a junction at stage t_i is succeeded by a junction at stage t_{i+1} is followed, then there are a finite number of overall paths and in principle it is trivial to select the one with minimum length. Practically, however, the number of paths escalates very quickly and even the most straightforward maps can lead to 10^6 or 10^7 paths all requiring the sum of the individual steps. The dynamic programming method can often reduce this work quite dramatically, for instance, in one well quoted example from 10^6 to 200 additions.

In practice most examples do not have this discrete nature and another well quoted example, that of a 400 m runner, describes this very well. Within his total energy capacity, which is supposed to be exhausted at the end of the race, the runner at each moment in time (i.e. continuous decision times) can decide to change his speed (within reasonable limits) to any other speed and so has to choose from an infinite number of speeds. While this illustrates the nature of a problem with continuous decisions and a continuum of choices, a possible mathematical *model* of such a runner would insist that the runner could only change his speed at two-second intervals and the choice would be to add -1.0, -0.5, 0, $+0.5$, $+1.0\ \text{m sec}^{-1}$ to his speed at each of these decision points. Again the problem is back to a discrete one of the type already described, with the additional complication of the constraint given by the total energy capacity.

One final comment in these introductory paragraphs is that in the problems considered here the choice and the costs are completely deterministic with no probabilistic element. Such stochastic problems are very important, particularly in business applications where random fluctuations appear continually in the calculations. These will be left for the reader to consult a more substantial book on dynamic programming such as Bellman and Dreyfus (1962), or Hastings (1973).

It is possible to write the above introductory paragraphs in a more mathematical manner. At a *stage* t_i the system under consideration can be in a finite number of *states* $\mathbf{x}_1, \mathbf{x}_2, \ldots, \mathbf{x}_M$, each of these states being specified uniquely by a P-vector. The cost of proceeding from state α at stage i to state β at stage $(i + 1)$ is given in the problem and can be written formally as $C(i, \mathbf{x}_\alpha; i + 1, \mathbf{x}_\beta)$. An *admissible* path for such a problem is defined to be a sequence of decisions at the successive stages $t_1, \ldots, t_N$ so that a decision at stage i leads to stage $(i + 1)$ and is allowable under the specification of the problem. The cost of an admissible path is just the sum of the C's along the path. The object of the problem is to select, out of all possible admissible paths, one which minimizes the overall cost.

To solve such problems by dynamic programming requires that the *principle of optimality* holds. The principle states that, if an optimal path from the initial stage t_1 to the final stage t_N exists and at an intermediate stage t_k it passes through the stage $\mathbf{x}_\alpha$, then the optimal path from $\mathbf{x}_\alpha$ at stage t_k to stage t_N coincides with the overall optimal path. Provided this principle is satisfied it is only necessary to formalize the idea to derive an algorithm that can be used to solve the problem.

Let $S(i, \mathbf{x}_\alpha)$ denote the cost of the *optimal* path from state $\mathbf{x}_\alpha$ at stage t_i to the final stage t_N. It is possible to proceed from state $\mathbf{x}_\alpha$ at stage t_i by taking an allowable step to $\mathbf{x}_\beta$ at stage t_{i+1}. The optimal path from $\mathbf{x}_\alpha$ at stage t_i must pass through one of the allowable states at stage t_{i+1}, say $\mathbf{x}_\gamma$. By the principle of optimality, the optimal path from $\mathbf{x}_\gamma$ at stage t_{i+1} to the final stage must coincide with the optimal path from $\mathbf{x}_\alpha$ at stage t_i. Thus comparing the sum of the costs, from $\mathbf{x}_\alpha$ to $\mathbf{x}_\beta$ and from $\mathbf{x}_\beta$ to the final stage, over *all* β will give the minimum cost.

$$S(i, \mathbf{x}_\alpha) = \min_{\text{all allowable } \beta} [C(i, \mathbf{x}_\alpha; i + 1, \mathbf{x}_\beta) + S(i + 1, \mathbf{x}_\beta)]. \tag{10.1}$$

The C's are all given in the problem so that provided all the S's are known at $(i + 1)$ it is possible to compute from (10.1) the S's at i for each state $\mathbf{x}_\alpha$. Assuming that the data at the final stage t_N is given then (10.1) can be used recursively to compute the S's at t_{N-1}, then at t_{N-2}, etc, until the required data at t_1 is obtained.

As an example consider the following problem. A Sheffield family wishing to go on holiday on the south coast of England work out the routes shown in figure 10.1. If the distance travelled is to be minimized, find to which resort the family should go and which route they should take. Abbreviating the towns by their first two letters, the path Sh → Le → Ox → Gu → Br is admissible while the path Sh → Le → Ox → Bi → Gl → Sa → Bo is not, since, in going from Oxford to Birmingham, the progression is from stage 2 to stage 1 and an admissible path insists stage 3 follows stage 2. Likewise the path Sh → Le → Gl → Ye → Ly is not admissible, since, on the map, the direct link Le → Gl is not allowed. For this problem it will be noted that there are 16 possible paths each requiring 4 additions, so evaluating all paths needs 64 additions. An N-stage problem of similar type would require $N \times 2^N$ additions.

The dynamic programming method starts with stage 3 and says that *if* the optimum path passes through Maidstone then the final step to stage 4 must be to Folkestone and of length 35 miles. Similarly *if* the optimum path passes through Guildford then the final step must be to Brighton of length 42 miles; likewise Sa → Bo length 30 miles and Ye → Do length 20 miles. The overall optimum path must pass through one of these four towns and hence the final step is one of the four indicated. Now look at stage 2: if the optimum path passes through London then the choices are (i) to Maidstone (36 miles) plus the minimum from Maidstone to the coast, which was found to be the 35 miles

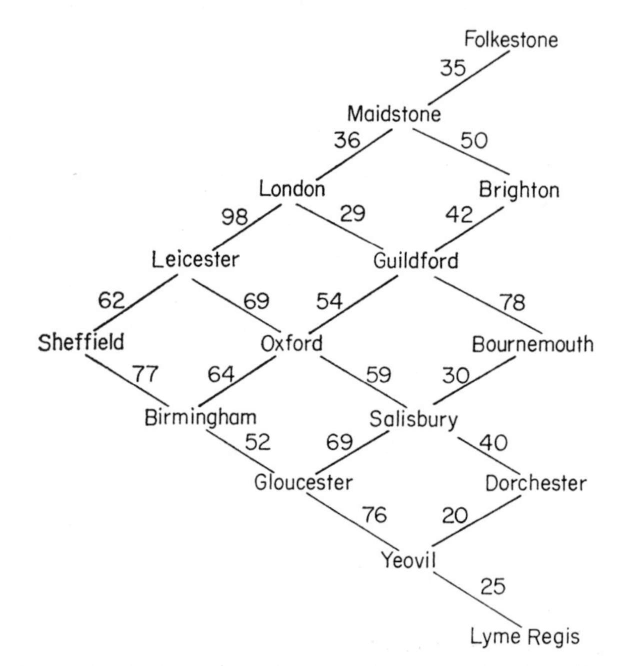

Figure 10.1 Chosen routes, with distances given in miles.

to Folkestone, giving a total of 71 miles; (ii) to Guildford (29 miles) plus the minimum from Guildford to the coast, that is to Brighton (42 miles), giving a total of 71 miles. Thus in this case there is nothing to choose between the two routes and either path is optimum for the journey from London to the coast. The towns Oxford and Gloucester can be treated similarly at stage 2, then Leicester and Birmingham at stage 1 and finally Sheffield at stage 0. This information is best presented as in figure 10.2, with the encircled numbers indicating the minimum distance from the town to the coast and the arrows denoting the optimum direction at each town. Following the arrows through gives the optimum path Sh → Le → Ox → Sa → Bo with a minimum distance of 220 miles. This particular problem has a unique solution but general problems of this sort may have several equally valid solutions.

It is interesting to compute the number of additions required in this method and then compare with the 64 required for the enumeration of all paths. The calculation at stage 3 requires no additions; at stage 2, 3×2; at stage 1, 2×2; and at stage 0, 2×1; giving a total of 12 additions, which is a saving of a factor of five. For an N-stage problem of this type the number of additions required is $2[(N-1) + (N-2) + \cdots + 1] = N(N-1)$, compared with $N \times 2^N$ for complete enumeration. Taking $N = 10$ this gives 90 and 10 240 respectively with a saving of effort of a factor of over 100; this changes the problem from a computer one to a hand computation problem.

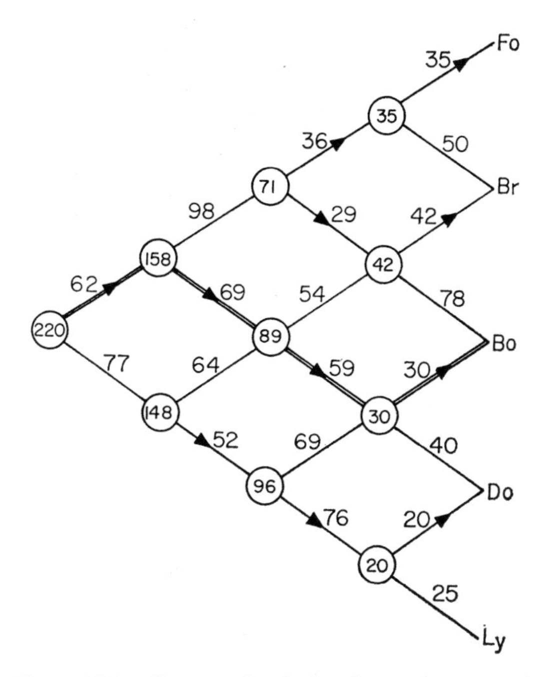

Figure 10.2 Computed solution for optimum route.

Although the above calculation shows an impressive saving of effort, the dynamic programming itself can run into serious computational difficulties for higher dimensional problems. Recall that the state α at stage i is specified by a P-vector, $\mathbf{x}_\alpha$. Suppose that in going from stage i to stage $(i + 1)$ each component of this P-vector can change independently to Q different values. Each application of (10.1) will therefore involve $k = P^Q$ additions. Taking quite modest values, $P = 4$, $Q = 5$ gives $k = 1\,028$. If at each stage the parameter α can take 10 values then the complete calculation at each stage requires about 10^4 additions. Thus quite small looking calculations can easily lead to major computational problems even using dynamic programming.

The formal algorithm (10.1) can now be illustrated on the problem studied above. The conventional method of treating such problems is to draw the map on a cartesian plane so that decision points have integer coordinates as

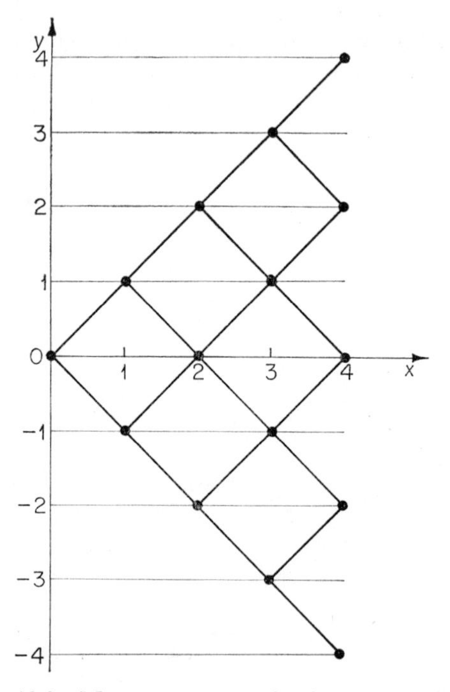

Figure 10.3 Map arrangement for formal algorithm.

in figure 10.3. The stages (t_i) are now specified by the x-values and the states ($\mathbf{x}_\alpha$) by the y-values. Thus the notation becomes somewhat simpler for this problem,

$$C(x, y; x + 1, z) = \text{distance from } (x, y) \text{ to } (x + 1, z)$$

and

$$S(x, y) = \text{optimal cost from } (x, y) \text{ to the line } x = 4,$$

so that the algorithm reads

$$S(x, y) = \min_{z=y\pm 1} [C(x, y; x + 1, z) + S(x + 1, z)].$$

The algorithm is applied by first noting that $S(4, z) = 0$ for all z, since this is just the distance of going to the coast if you are already there. Thus putting $x = 3$ into the above enables $S(3, y)$ to be evaluated for $y = -3, -1, +1, +3$.

Since $S(3, y)$ is available for all y, put $x = 2$ into the above; all the values in the right-hand side are known so $S(2, y)$ can be computed for $y = -2$, 0, $+2$. Continuing in this way $S(0, 0)$ eventually can be found.

In application devising an appropriate dynamic programming form is not always easy. It often requires considerable ingenuity, although once done it is usually obvious. Provided, however, the principle of optimality holds the problem can be forced into a dynamic programming form.

10.2 Transportation and Scheduling Problems

These two types of problem provide the commonest applications of discrete dynamic programming. The problem in section 10.1 was a typical transportation problem. Another similar example will now be quoted to illustrate one or two slightly different points.

Example 63 Find the route which minimizes the distance between Glasgow and London using the road mesh shown in figure 10.4. (Those familiar with the British road system will recognize the main trunk routes.)

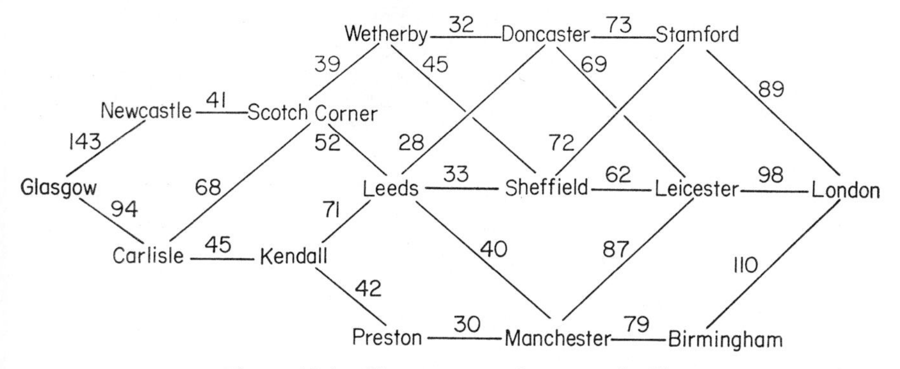

Figure 10.4 Chosen routes for example 65.

The method of working is identical with the example in section 10.1 and it is left to the reader to show that the route with optimum distance is Glasgow–Carlisle–Scotch Corner–Wetherby–Doncaster–Stamford–London with a total distance of 395 miles. The main interest in the problem is to compute the formal algorithm. This is not so straightforward since the possible choices at each decision point can be one, two or three and *not* as in the previous example a uniform two. The best method of dealing with this situation is to make the mesh uniform by inserting fictitious roads between places with *infinite* length to ensure that they are not chosen. The present example can be modelled with towns on points with integer coordinates as shown in figure 10.5. It is left to the reader to complete the formal algorithm in terms of

$$C(x, y; x + 1, z) = \text{cost from } (x, y) \text{ to } (x + 1, z)$$

and

$$S(x, y) = \text{optimal cost from } (x, y) \text{ to } (6, 0).$$

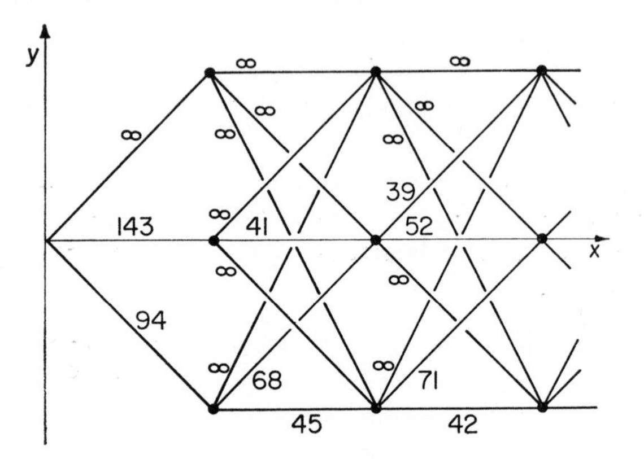

Figure 10.5 Map arrangement for formal algorithm, example 63.

A scheduling problem of some interest is as follows. (Incidentally it should be noted that this problem involves a constraint and also it can be solved by integer linear programming, section 3.4.)

Example 64 A ship can carry 11 units of weight. Four possible types of object can be transported each having given weight and value to the shipper as shown in table 10.1. Find the maximum possible value of the cargo.

Table 10.1 Details of cargo to be transported.

Type (i)	weight (w_i)	value (v_i)
1	4	7
2	3	6
3	2	3
4	1	2

Suppose the ship is loaded with N_1 of type 1, N_2 of type 2, . . . then to satisfy the weight constraint

$$\sum_{i=1}^{4} N_i w_i \leq 11, \tag{10.2}$$

while the value of the cargo is

$$V = \sum_{i=1}^{4} N_i v_i. \tag{10.3}$$

It is required therefore to maximize V from (10.3) subject to (10.2) over all possible integers $N_i = 0, 1, 2, \ldots$. Let

$$f_k(z) = \text{maximum value of load of weight } z \text{ over all } N_1, N_2, \ldots, N_k$$

$$\text{subject to} \sum_{i=1}^{k} N_i w_i \leq z;$$

this optimally fills the ship up to a weight z with the first k type of object. Thus the above definition implies that the stages are labelled as k, i.e. how many types that have been loaded, and the state is measured by z, i.e. the weight to which the ship can be loaded. But $f_1(z)$ just loads type 1, so immediately $f_1(z) = 0$ if $z < 4$, 7 if $4 \leq z < 8$, 14 if $8 \leq z \leq 11$. Now try to compute $f_2(z)$ by finding the maximum value by comparing the different possible cases for each z; thus put in zero of type 2 and fill up optimally to weight z with type 1 or put in one of type 2 and fill up optimally to weight $(z - 3)$ with type 1, or put in two of type 2 and fill up optimally to weight $(z - 6)$ with type 1, etc. Choose the maximum value out of all these cases and repeat for each z, thus $f_2(z)$ is now known for all z. To compute $f_3(z)$ the process is similar: try zero of type 3 and fill optimally to weight z with types 1 and 2, i.e. $f_2(z)$, or try one of type 3 and fill optimally to weight $(z - 2)$ with types 1 and 2, i.e. $f_2(z - 2)$, etc. Again the calculation is performed recursively and can be written generally as

$$f_k(z) = \underset{\substack{N = 0, 1, 2, \ldots \\ \text{with} \\ (z - Nw_k) \geqslant 0}}{\text{maximum}} [f_{k-1}(z - Nw_k) + Nv_k], \tag{10.4}$$

the inequality constraint ensures that the ship is carrying a positive or zero weight. The boundary condition is $f_0(z) = 0$ for all z. $f_1(z)$ can then be calculated from (10.4) for all z; then $f_2(z)$; then $f_3(z)$, etc; eventually, for this particular problem, $f_4(11)$ is required. The results are best presented in tabular form (see table 10.2).

Table 10.2 Dynamic programming solution using (10.4).

z	0	1	2	3	4	5	6	7	8	9	10	11
$f_1(z)$	0	0	0	0	7	7	7	7	14	14	14	14
$f_2(z)$	0	0	0	6	7	7	12	13	14	18	19	19
$f_3(z)$	0	0	3	6	7	9	12	13	15	18	19	21
$f_4(z)$												22

If the optimal decision is recorded at each calculation of $f_k(z)$, the maximum value of 22 can be found to be made up of $(0v_1 + 0v_2 + 0v_3 + 11v_4)$ or $(0v_1 + 1v_2 + 0v_3 + 8v_4)$ or $(0v_1 + 2v_2 + 0v_3 + 5v_4)$ or $(0v_1 + 3v_2 + 0v_3 + 2v_4)$. This gives four possible optimal solutions calculated without the enormous effort of enumerating all possible loadings.

There is a wide range of transportation and scheduling problems that can be treated in this way and the problems at the end of the chapter provide further examples.

10.3* Continuous Case

In section 10.1 the problem of the 400 m runner showed how a continuous problem could be formulated having a continuum of decision points and a continuum of choices. It is not at all obvious mathematically that such a problem has a solution although intuitively one would be expected. At least in the discrete case there are a finite (albeit extremely large) number of admissible paths, each with a calculated cost, and the minimum cost can be chosen from this finite set. For an infinite set it is now no longer obvious that a minimum belongs to the set. Even if this minimum does exist continuous problems can rarely be solved analytically and a numerical procedure is normally used. This involves a discretization, as described for the 400 m runner, and the problem is now back to the much simpler discrete problem of section 10.1.

One way of introducing the continuous case is to look at the classical variational problem. Take, for instance, the problem of minimizing the cost functional

$$I[y] = \int_{t_1}^{t_2} f(t, y, \dot{y})\, dt \tag{10.5}$$

with

$$y(t_1) = A, \qquad y(t_2) = B,$$

and $y(t) \in C^2$. The dynamic programming approach considers the optimum path from T to t_2 as

$$S(T, y) = \min_{z} \int_{T}^{t_2} f(t, z, \dot{z})\, dt \tag{10.6}$$

with

$$z(T) = y, \qquad z(t_2) = B.$$

In this formulation it should be noted that the stage is defined by T, which is a continuous variable, and the state by y, which can take a continuum of values.

Now this is made up of the cost of getting from y at T to $(y + \delta y)$ at $(T + \delta T)$ and cost of the optimum path from $(y + \delta y)$ at $(T + \delta T)$ to the end; or mathematically

$$S(T, y) = \min_{z} \left(\int_{T}^{T+\delta T} f(t, z, \dot{z})\, dt + S(T + \delta T, y + \delta y) \right) \tag{10.7}$$

with

$$z(T) = y, \qquad z(T + \delta T) = y + \delta y.$$

Assuming δT to be small then $\delta y = \dot{y}\, \delta T$, the integral in (10.7) can be written $\delta T f(T, y, \dot{y})$ to $0(\delta T)$. The minimization is now over δy; this in turn means over $\dot{y}$. Hence using these results together with Taylor's theorem

$$S(T, y) = \min_{\dot{y}} \left(\delta T f(T, y, \dot{y}) + S(T, y) + \frac{\partial S}{\partial T} \delta T + \frac{\partial S}{\partial y} \dot{y}\, \delta T + 0(\delta T^2) \right).$$

But the terms $S(T, y)$ and $(\partial S/\partial T)\, \delta T$ are independent of $\dot{y}$ and hence in the limit $\delta T \to 0$

$$-\frac{\partial S}{\partial T} = \min_{\dot{y}} \left(\frac{\partial S}{\partial y} \dot{y} + f(T, y, \dot{y}) \right) \qquad \text{(Bellman's equation)}. \tag{10.8}$$

This is the basic recurrence relation for the functional chosen in (10.5) and is much more formidable than the discrete case. However, minimizing the right-hand side of (10.8) formally with respect to $\dot{y}$ gives at the optimum $y = Y$

$$0 = \frac{\partial S}{\partial Y} + \frac{\partial f}{\partial \dot{Y}}(T, Y, \dot{Y}), \tag{10.9}$$

and at the optimum (10.8) just becomes

$$-\frac{\partial S}{\partial T} = \frac{\partial S}{\partial Y} \dot{Y} + f(T, Y, \dot{Y}). \tag{10.10}$$

Now differentiate (10.9) with respect to T to give

$$\begin{aligned} 0 &= \frac{d}{dT}\left(\frac{\partial S}{\partial Y}\right) + \frac{d}{dT}\left(\frac{\partial f}{\partial \dot{Y}}\right) \\ &= \frac{\partial}{\partial Y}\left(\frac{\partial S}{\partial T} + \frac{\partial S}{\partial Y} \dot{Y}\right) + \frac{d}{dT}\left(\frac{\partial f}{\partial \dot{Y}}\right). \end{aligned}$$

But using (10.10) on the first term of this equation produces the familiar equation

$$0 = -\frac{\partial f}{\partial Y} + \frac{d}{dT}\left(\frac{\partial f}{\partial \dot{Y}}\right).$$

Indeed this is just the Euler–Lagrange equation so that, provided an optimum solution exists for (10.5), the necessary condition for an optimum will be obtained from (10.8) and a useful numerical process can be constructed from a discrete form of (10.7).

Example 65 Use a dynamic programming method to solve the brachistochrone problem (section 7.1) discretizing the problem suitably and using a version of (10.7).

It will be recalled that the functional for the descent time is

$$I[y] = \int_0^a \left(\frac{1 + y'^2}{y}\right)^{\frac{1}{2}} dx$$

with

$$y(0) = 0, \qquad y(a) = b.$$

Take for convenience $a = b = 4$ as shown in figure 10.6. Letting

$$S(x, y) = \text{path from } (x, y) \text{ to } (4, 4) \text{ giving minimum time}$$

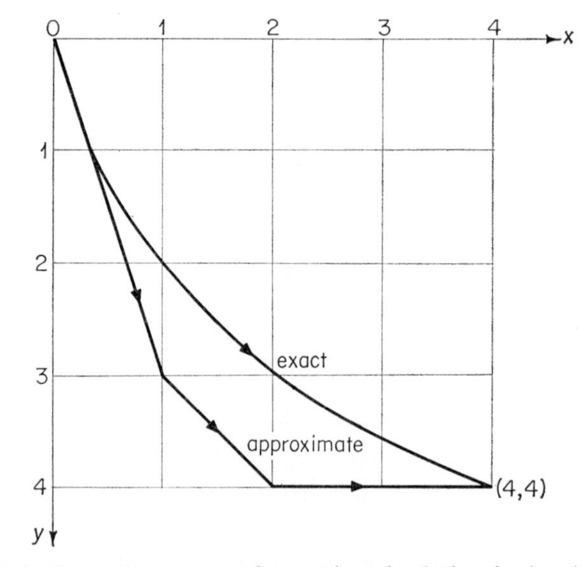

Figure 10.6 Dynamic programming mesh and solution for brachistochrone.

then (10.7) for the present case becomes

$$S(x, y) = \min_{\delta y} \left[S(x + \delta x, y + \delta y) + \int_x^{x+\delta x} \left(\frac{1 + y'^2}{y} \right)^{\frac{1}{2}} dx \right].$$

Assuming that, in the interval δx, y' is a constant so that $\delta y = y' \, \delta x$ and taking $\bar{y} = y + \frac{1}{2}\delta y$ in this interval the equation becomes

$$S(x, y) = \min_{\delta y} \left[S(x + \delta x, y + \delta y) + \delta x \left(\frac{1 + (\delta y/\delta x)^2}{\bar{y}} \right)^{\frac{1}{2}} \right]. \quad (10.11)$$

It now remains to choose the mesh, compute the final term of (10.11) and then use the recurrence relation defined by (10.11) to compute $S(0, 0)$, starting with $S(4, 4) = 0$. Take $\delta x = 1$ and $\delta y = 0, 1, 2, 3, 4$ then the values of $H(y, \delta y) = [1 + (\delta y/\delta x)^2/\bar{y}]^{\frac{1}{2}}$ are given in table 10.3.

Table 10.3 Values of $H(y, \delta y)$, example 65.

y	δy 0	1	2	3	4
0	∞	2	2.236	2.582	2.916
1	1	1.155	1.581	2.000	
2	0.707 1	0.894 4	1.291		
3	0.577 4	0.755 9			
4	0.500 0				

The algorithm proceeds from $x = 3$ where there is no choice of path since the final step must be to (4, 4)

$$S(3, 4) = \min [S(4, 4) + H(4, 0)] = 0.500\,0$$

$$S(3, 3) = \min [S(4, 4) + H(3, 1)] = 0.755\,9$$

and similarly $S(3, 2) = 1.291$, $S(3, 1) = 2.000$, $S(3, 0) = 2.916$.

Now take $x = 2$

$$S(2, 4) = \min [S(3, 4) + H(4, 0)] = 1.000$$

$$S(2, 3) = \min \begin{bmatrix} S(3, 3) + H(3, 0) = 1.333 \\ S(3, 4) + H(3, 1) = 1.256 \end{bmatrix} = 1.256$$

$$S(2, 2) = \min \begin{bmatrix} S(3, 2) + H(2, 0) = 1.998 \\ S(3, 3) + H(2, 1) = 1.650 \\ S(3, 4) + H(2, 2) = 1.791 \end{bmatrix} = 1.650$$

and similarly $S(2, 1) = 2.337$, $S(2, 0) = 3.338$. The $S(1, y)$ can be calculated in terms of the now known $S(2, y)$ and finally the required result is $S(0, 0) = 4.338$. Working back through the calculation the optimum path is $(0, 0) \to (1, 3) \to (2, 4) \to (3, 4) \to (4, 4)$ which is illustrated in figure 10.6. These results should be compared with the exact result of $S(0, 0) = 5.16$ and the exact curve also shown in the figure. A finer mesh would, of course, give a more accurate result but the calculation was performed largely to illustrate the use of the dynamic programming method in a variational problem. The discretized recurrence relation (10.11) in the notation used becomes

$$S(x, y) = \min_{\substack{N=0,1,2,\ldots \\ (y+N)\leqslant 4}} [S(x+1, y+N) + H(x, N)].$$

10.4* Differential Equation Constraints

It was fortunate that in the problem selected in section 10.3 the equation of motion could be solved exactly and hence the cost functional evaluated to reduce the problem to a standard variational form. In many examples, however, this exact solution is quite out of the question. To identify the difficulty consider the problem of finding the function $u(t)$ which minimizes the cost functional

$$I[u] = \int_0^T f(t, x, \dot{x}, u)\, dt \tag{10.12}$$

where

$$\dot{x} = g(t, x, u) \tag{10.13}$$

with

$$x(0) = 1.$$

For each $u(t)$ the differential equation (10.13) can be solved (at least in principle), subject to the given boundary condition, to give the function $x(t)$; these $x(t)$, $u(t)$ can then be substituted back into (10.12) to give the cost. This method can be followed for each $u(t)$, the costs compared and the minimum chosen. It is precisely this sort of problem that occurs in control theory and in the next chapter such problems will be dealt with in more detail. The present analysis just attempts to show how a simple example of this type can be treated by a dynamic programming technique to produce a useful numerical procedure. The fact that a numerical method is derived is no real loss, since the solution of (10.13) for a given $u(t)$ would almost certainly require a numerical solution which must then be put into (10.12). Such procedures are quite lengthy and are usually beyond a hand computation.

Consider the particular example of minimizing

$$\int_0^1 (10x^2 + u^2)\, dt \tag{10.14}$$

subject to

$$\dot{x} + 2x = u \tag{10.15}$$

with

$$x(0) = 1.$$

The functional contains the terms x^2 and u^2 to prevent either getting too large. The first step in the calculation is to divide the time interval [0, 1] into N equal parts $0 = t_0 < t_1 < t_2 < \cdots < t_N = 1$, with $t_{n+1} - t_n = \Delta t$ and to let x_n, u_n denote the values of $x(t)$ and $u(t)$ at $t = t_n$. The differential equation (10.15) can now be put into a discrete form

$$x_{n+1} = (1 - 2\Delta t)x_n + \Delta t u_n, \tag{10.16}$$

with

$$x_0 = 1.$$

More sophisticated versions of this equation are possible, for instance by solving (10.15) exactly on the assumption that $u = u_n$ over the interval Δt. This formula will be satisfactory for the present purposes, however, since it is the method that is important, accuracy being a secondary consideration. Having written (10.15) in discrete form, the functional is rewritten

$$\sum_{n=0}^{N-1} (10x_n^2 + u_n^2) \tag{10.17}$$

and the minimization must now be performed over all sets $(u_0, u_1, \ldots, u_{N-1})$. The argument used is now the dynamic programming one, starting with $k = N$ and working backwards to $k = 0$. Let

$$S_k(x_k) = \min_{u_k, \ldots, u_{N-1}} \sum_{n=k}^{N-1} (10x_n^2 + u_n^2),$$

where, for a given x_k and $u_k, u_{k+1}, \ldots, u_{N-1}$, the values $x_{k+1}, \ldots$ are obtained from (10.16); this expression just calculates the optimum path over last steps from k to N starting at a value x_k. Suppose $S_{k+1}(x_{k+1})$ has been calculated for *all* x_{k+1} then the above can be written

$$S_k(x_k) = \min_{u_k, \ldots, u_{N-1}} \left(10x_k^2 + u_k^2 + \sum_{n=k+1}^{N-1} (10x_n^2 + u_n^2)\right).$$

But the minimization can be performed over $u_{k+1}, \ldots, u_{N-1}$ as

$$S_k(x_k) = \min_{u_k} [10x_k^2 + u_k^2 + S_{k+1}(x_{k+1})]$$

and hence using (10.16) to relate x_{k+1} to x_k and u_k, the final working equation is

$$S_k(x_k) = \min_{u_k} \{10x_k^2 + u_k^2 + S_{k+1}[(1 - 2\Delta t)x_k + \Delta t u_k]\}. \quad (10.18)$$

The method of working can now be summarized. At step $(k + 1)$, $S_{k+1}(x)$ is known for all x (in practice it will be known at a discrete set of values and intermediate values will be obtained by interpolation). Choose a particular value for x_k, then the right-hand side of (10.18) can be computed for each u_k since the first two terms are simple functions and S_{k+1} is known for all arguments. As u_k is varied the minimum for S_k is selected and the corresponding value of u_k (say u^*_k) at the minimum is noted. This minimization is performed for each of a mesh of values of x_k so that S_k and u^*_k are known at all the mesh points. Hence all the information is now known at step k and the process can then be repeated for $(k - 1)$. In this manner (10.18) is used recursively until $S_0(x_0) = S_0(1)$ is calculated and the discretized form of the optimal $u(t)$, $u^*_0, u^*_1, \ldots, u^*_{N-1}$ is obtained from working back through the recorded values.

The major step in the calculation is the minimization in (10.18) for each x_k. In practice this could be required at 20 values of x_k for each of a 10 time step calculation giving a total of 200 minimizations. For larger problems, serious storage problems can also be encountered, although these have become less acute as the size of digital computers has grown. An example of this would be seen if there were three equations in (10.15) for say x, y, z. At each of the 10 time steps, 20 values of x, y and z would be required, so that $20 \times 20 \times 20 \times 10 = 80\,000$ minimizations would be necessary and the storage of S_k and u^*_k alone would require 160*K*.

In the example quoted, all the minimizations can be done analytically because of the simplicity of the problem but it is instructive to perform a few steps of the calculation to illustrate the general method more clearly. Let $\Delta t = 0.1$ so that $N = 10$. The cost of going from step 10 to step 10 is, of course, zero, so the boundary condition becomes $S_{10}(z) = 0$ for all z. For $k = 9$ the equation (10.18) becomes, therefore

$$S_9(x_9) = \min_{u_9} (10x_9^2 + u_9^2) = 10x_9^2,$$

since the minimization is obvious with $u^*_9 = 0$ for all x_9. Now take $k = 8$ then

$$S_8(x_8) = \min_{u_8} [10x_8^2 + u_8^2 + S_9(0.8x_8 + 0.1u_8)]$$

$$= \min_{u_8} [10x_8^2 + u_8^2 + 10(0.8x_8 + 0.1u_8)^2].$$

Table 10.4 Optimum solution for u and x.

k	0	1	2	3	4	5	6	7	8	9	10
α_k	1.397	1.394	1.389	1.383	1.362	1.339	1.263	1.093	0.727 3	0	–
β_k	21.18	21.16	21.11	21.00	20.89	20.51	20.10	18.74	15.82	10	0
x_k	1	0.660 3	0.436 2	0.281 5	0.186 4	0.123 7	0.082 4	0.055 5	0.038 4	0.027 9	0.022 3
u_k^*	−1.397	−0.920 5	−0.605 8	−0.389 6	−0.253 9	−0.165 7	−0.104 1	−0.060 7	−0.027 9	0	–

This minimization can also be performed for given x_8 by standard calculus to give

$$2u_8 + 2(0.8x_8 + 0.1u_8) = 0$$

or

$$u^*_8 = -0.727\,3x_8$$

and

$$S_8(x_8) = 15.82x_8^{\,2}.$$

The calculations of $S_7(x_7)$, $S_6(x_6)$ . . . follow similarly until $S_0(1)$ is obtained. The simplifying feature of this calculation is that all the minimizations lead to $u^*_k = -\alpha_k x_k$ and $S_k(x_k) = \beta_k x_k^{\,2}$, where α_k and β_k are constants that are calculated as above.

The values obtained are given in table 10.4 and $S_0(1) = 21.18$ gives the minimum cost.

This particular example gives a clear picture of how the calculation proceeds for any problem of the type described by (10.12) and (10.13). It can be appreciated how much work is involved if the minimization has to be done numerically, although it should be noted that any other method would almost certainly require as much if not more work.

An important modification of this type of method occurs if a *constraint* is imposed on u, say $|u| \leq 1$. It was emphasized in Chapter 1 how frequently constraints operate so it is crucial that methods should adapt easily to them. In the present case the basic algorithm is similar; it is necessary, however, to plug in an alternative minimization routine that deals satisfactorily with the constraints and to take great care with the discontinuities that can arise. It is here that the major weakness of the dynamic programming method appears.

Another important modification is to the case where the functional is governed by a *set* of differential equations (see problem 6 and chapter 11). Again the basic algorithm is precisely the same although the size of the calculation grows enormously.

PROBLEMS

1. Given the array of numbers

$$\begin{array}{ccccc} 7 & 7 & 6 & 2 & 8 \\ 3 & 8 & 2 & 6 & 6 \\ 8 & 6 & 4 & 2 & 0, \\ 3 & 6 & 6 & 3 & 6 \\ 1 & 3 & 7 & 4 & 8 \end{array}$$

use a dynamic programming method to find the path from the top left-hand corner to the bottom right-hand corner that minimizes the sum of the numbers along the path. An allowable step is a move to the number immediately to the right or immediately below.

Write a formal dynamic programming algorithm for this problem.

2. The cost of taking an admissible step between nodes from left to right is shown in figure 10.7. Find the path from A to B which gives minimum cost. Write a formal algorithm to solve this problem.

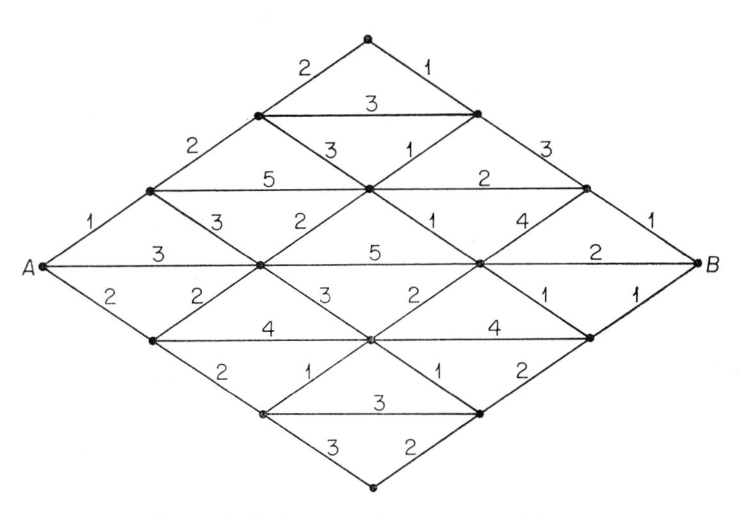

Figure 10.7 Mesh and costs for problem 2.

3. The cost of replacing a machine is 4 units. The profit in a year from a machine that is 0, 1, 2, 3 years old (at the start of the year) is 6, 5, 3, 1 respectively. The machine must be scrapped at the end of the fourth year. Find the replacement policy over a six-year period which maximizes the overall profit starting with a new machine (use $f(N, M)$ = the maximum profit from the optimal policy if currently the machine is N years old and there are M years of the period to run).

4. A job consists of producing a specified number of items. The work is done in 5 batches and the cost, $g(N, M)$, of producing M items in the Nth batch is given in table 10.5; note that the cost of lying idle is also included.

If $f(N, P)$ is the minimum cost to produce a *total* of P items in the *Nth or later* batches, show that

$$f(N, P) = \underset{I=0,1,2,3,4}{\text{minimum}} \, [f(N+1, P-I) + g(N, I)],$$

where $f(N+1, P-I)$ is given the value 1 000 if $(P-I)$ is negative.

Find the minimum cost of producing 10 items, i.e. find $f(1, 10)$.

Table 10.5 Given costs for problem 4.

Number of items produced	Batch 1	2	3	4	5
0	2	2	3	5	4
1	3	3	4	6	8
2	7	6	8	8	10
3	10	11	13	16	15
4	11	12	14	20	17

5. Use the dynamic programming method of section 10.3, with a suitable mesh, to find the minimum of the functional

$$\int_0^1 (y'^2 - y^2)\, dx, \qquad y(0) = 0,\ y(1) = 1.$$

Compare your result with the exact answer obtained from solving the Euler equation.

6. Show that an entirely analogous result to (10.18) can be set up for the general functional (10.12) subject to (10.13) and the given boundary condition. Similarly deal with the vector case of minimizing

$$I[\mathbf{u}] = \int_0^T f(t, \mathbf{x}, \dot{\mathbf{x}}, \mathbf{u})\, dt$$

with

$$\dot{\mathbf{x}} = \mathbf{g}(t, \mathbf{x}, \mathbf{u}) \quad \text{and} \quad \mathbf{x}(0) \text{ given.}$$

7. Solve (10.15) explicitly in the interval $t_n \leq t \leq t_{n+1}$ on the assumption that $u = u_n =$ constant in the interval and with the initial condition that $x = x_n$ when $t = t_n$. See how much effect this has on the first two steps of the 10-step process used in the text.

Chapter XI*

Control Problems

11.1* Introduction

In various previous chapters control-type problems were mentioned and it was left to the present chapter to collect this material together. All the quoted examples belong to *optimal control theory*, where some control variable is sought to minimize a given cost functional. The more traditional automatic control theory, involving largely stability considerations by Laplace transform techniques, will not be considered here.

A typical example of an optimal control mechanism is a rider on a bicycle. The rider travels with a constant speed in a straight line path and for some reason he deviates from this path. He then adjusts the handlebars of the cycle to bring himself back to his steady path as quickly as possible. From the mechanics of the bicycle, there is a complicated set of differential equations governing its motion, these incorporate an external or control force provided by the rider on the front wheel via the handlebars. The object of the rider is to adjust the angle of the front wheels, $\theta(t)$, to right himself as quickly as possible. Thus out of all possible adjustments the rider selects the one, $\bar{\theta}(t)$, which minimizes the overall time it takes to bring the cycle to its steady path. Taught by a little experience, the human brain easily provides the solution to this very complicated optimal control problem. However, if the controlling is to be done automatically by a computer, there is the solution of the governing differential equations to be obtained and an optimizing routine to be followed.

The above example is very complicated and is not convenient for illustrating the mathematics used. A more suitable simple example was provided in section 10.4. Find the minimum of

$$J[u] = \int_0^1 (10x^2 + u^2)\, dt \tag{11.1}$$

subject to

$$\dot{x} = -2x + u, \tag{11.2}$$

with the boundary condition $x(0) = 1$. Here there is a single control variable u and a single differential equation (11.2) to solve. The method of analysing this problem in section 10.4 was that of dynamic programming. An alternative approach will be given in section 11.2 where an almost identical problem will be solved.

In many modern problems a major additional difficulty is introduced by constraining the control variables. An example that has already been quoted illustrates this type of problem.

Example 66 Find the forcing function $F(t)$ required to minimize the time taken to move a body from rest at $x = a$ to rest at the origin, with

$$\ddot{x} = F \quad \text{and} \quad |F| \leq 1.$$

This problem can be written in a notation similar to that used above as: find the minimum of

$$I[F] = T \tag{11.3}$$

subject to

$$\dot{x}_1 = x_2, \qquad \dot{x}_2 = F, \tag{11.4}$$

with boundary conditions

$$x_1(0) = a, \qquad x_2(0) = 0, \qquad x_1(T) = 0, \qquad x_2(T) = 0,$$

and with the constraint on the control parameter

$$|F| \leq 1. \tag{11.5}$$

Several points should be noted from these examples which allow generalizations of the problem. Firstly the equations of motion (11.4) are written as a *set* of first order equations. In general these can be written in a convenient vector form in terms of $\mathbf{x} = (x_1, x_2, \ldots, x_n)$ as

$$\dot{\mathbf{x}} = \mathbf{f}(\mathbf{x}, \mathbf{u}, t), \tag{11.6}$$

where instead of a single control variable, a set of control functions $\mathbf{u} = (u_1, u_2, \ldots, u_n)$ is used. The boundary conditions on $\mathbf{x}$ are usually of two point type, some conditions on the x_i at $t = 0$ and some at $t = T$. The functional itself can also be generalized to a form which covers the majority of interesting cases

$$J[\mathbf{u}] = \theta[\mathbf{x}(t), t]\Big|_0^T + \int_0^T \varphi[\mathbf{x}(t), \mathbf{u}(t), t]\, dt, \tag{11.7}$$

where θ and φ are known functions, with φ usually a positive definite function. The final time, T, may be known, as in (11.1) with $T = 1$, or unspecified as in (11.3). In either case it is possible to specify various final (or initial) conditions but a very common type is to insist that at $t = T$

$$N[\mathbf{x}(T), T] = 0; \tag{11.8}$$

this is usually termed the *terminal manifold*. The two examples quoted give particularly simple forms for this terminal manifold. Finally it should be noted that the major difference between the two quoted examples was the inclusion of the *constraint* (11.5). This can also be generalized to the form

$$g(\mathbf{x}, \mathbf{u}, t) \geq 0. \tag{11.9}$$

Thus the most general problem that is normally considered is to minimize (11.7), subject to the set of differential equations (11.6), boundary conditions (11.8) and constraint (11.9). These constraints severely restrict the functions that are eligible as candidates for the optimum. Hence this optimum is normally dominated by the constraints and in such cases the solution is frequently *discontinuous*. This immediately puts traditional calculus methods into difficulties and methods must be devised to cope adequately with such discontinuities.

11.2* Problems With Unconstrained Control Parameters

Consider the single variable case of the type shown in (11.1) and (11.2); a fairly general form is to minimize

$$J[u] = \int_0^1 \varphi(x, u, t)\, dt \tag{11.10}$$

subject to

$$\dot{x} = f(x, u, t) \tag{11.11}$$

and boundary condition $x(0) = a$. In this case the time interval is fixed as $0 \leq t \leq 1$ but problems involving a variable final time will be considered later.

The Lagrange multiplier technique of section 6.4 is applicable, so that the unconstrained optimum of

$$\begin{aligned} J^*[u, x] &= \int_0^1 [\varphi + \lambda(t)(f - \dot{x})]\, dt \\ &= -[\lambda x]_0{}^1 + \int_0^1 [(\varphi + \lambda f) + \dot{\lambda} x]\, dt, \end{aligned}$$

is required. Defining

$$H(x, u, t) = \varphi + \lambda f \tag{11.12}$$

the Euler equations for the functional, with integrand $F = H + \lambda x$, are just

$$0 = \frac{d}{dt}\left(\frac{\partial F}{\partial \dot{u}}\right) - \frac{\partial F}{\partial u} = \frac{\partial H}{\partial u}$$

$$0 = \frac{d}{dt}\left(\frac{\partial F}{\partial \dot{x}}\right) - \frac{\partial F}{\partial x} = -\left(\frac{\partial H}{\partial x} + \dot{\lambda}\right).$$

Thus

$$0 = \frac{\partial H}{\partial u}, \qquad \dot{\lambda} = -\frac{\partial H}{\partial x} \tag{11.13}$$

and from (11.11) and (11.12)

$$\dot{x} = f = \frac{\partial H}{\partial \lambda}. \tag{11.14}$$

The boundary conditions at $t = 1$ must be computed from first principles by putting $x = X + \varepsilon\alpha(t)$, $u = U + \delta\beta(t)$, where X, U give the optimum and to satisfy the given boundary condition $\alpha(0) = 0$. The change in J^* from the optimum value is given to first order in ε, δ as

$$\begin{aligned} \Delta J^* &= -\varepsilon[\lambda\alpha]_0^1 + \int_0^1 (H\beta\delta + H_x\alpha\varepsilon + \dot{\lambda}\alpha\varepsilon)\, dt \\ &= -\varepsilon\lambda(1)\alpha(1) + \delta\int_0^1 (\beta H_u)\, dt + \varepsilon\int_0^1 \alpha(H_x + \dot{\lambda})\, dt. \end{aligned}$$

Applying the usual variational argument that $\varepsilon = \delta = 0$ gives the extremum, the two integral terms in the above lead to equations (11.13). The natural boundary conditions now give $\lambda(1)\alpha(1) = 0$, but since $\alpha(1)$ was left unspecified, $\lambda(1) = 0$ and hence the conditions at the two ends are

$$x(0) = a, \qquad \lambda(1) = 0.$$

The function u, in the above, is selected to minimize H, since $\partial H/\partial u = 0$ from (11.13). This is the *Pontryagin minimum principle* (maximum if H is defined as $H = -\varphi + \lambda f$) in its simplest context. It will be seen that it is this result which can be generalized to the more complicated constrained situations. Those familiar with classical mechanics (cf. problem 9, p. 123) will recognize the equation in (11.13) and (11.14) as the standard Hamiltonian equations of motion, and for this reason H is referred to as the Hamiltonian.

Take as an example a problem similar to (11.1), (11.2):

Example 67 Minimize

$$J[u] = \int_0^1 (12x^2 + u^2)\, dt$$

subject to

$$\dot{x} = u - 2x \qquad \text{and} \qquad x(0) = 1.$$

Now the Hamiltonian is

$$H = 12x^2 + u^2 + \lambda(u - 2x)$$

and the canonical equations are

$$H_u = 0 = 2u + \lambda, \qquad H_x = 24x - 2\lambda = -\dot{\lambda}$$

together with

$$\dot{x} = u - 2x$$

and boundary conditions $x(0) = 1$, $\lambda(1) = 0$. Eliminating the control variable u, the resulting pair of equations,

$$\dot{\lambda} = 2\lambda - 24x, \qquad \dot{x} = -\tfrac{1}{2}\lambda - 2x,$$

are linear and can be solved easily. The major difficulty is the *two* point nature of the boundary conditions; in the present case the problem is not too hard but for more complicated cases it is here that the major computational problems arise. Integrating gives, after a little work

$$u = -\tfrac{1}{2}\lambda = -6\,\frac{\sinh 4(1 - t)}{\sinh 4 + 2\cosh 4}$$

$$x = \frac{\sinh 4(1 - t) + 2\cosh 4(1 - t)}{\sinh 4 + 2\cosh 4},$$

and these curves are plotted in figure 11.1.

As a final example in this section consider a problem that includes one or two of generalizations suggested in section 11.1.

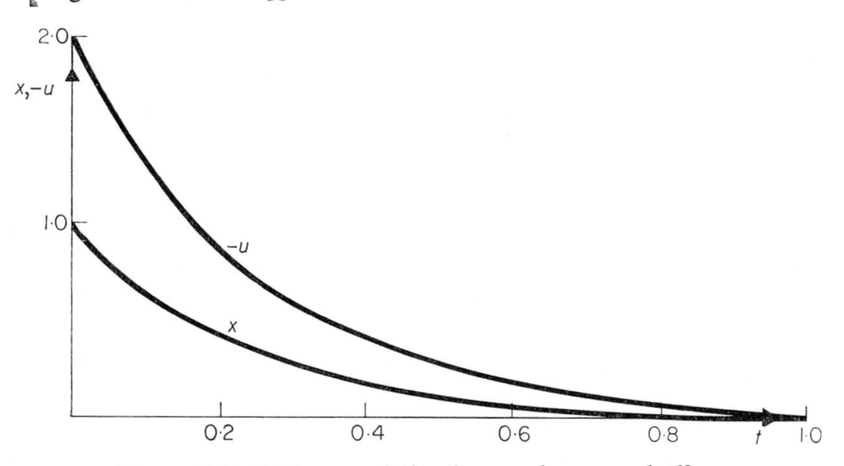

Figure 11.1 Optimum solution for x and u, example 69.

Example 68 Minimize

$$J[u, T] = T^2 + \int_0^T (x_1{}^2 + u^2)\, dt$$

with

$$\ddot{x}_1 = u$$

and $x_1(0) = 1$, $\dot{x}_1(0) = 0$, $x_1(T) = 0$, $\dot{x}_1(T)$ unspecified.

Physically this problem corresponds to minimizing the given cost functional when an object is moved, under a force per unit mass of $u(t)$, from rest at $x_1 = 1$ to the origin. The time taken and the final speed are left unspecified. It is not difficult to imagine similar problems occurring in the trajectories of artificial satellites with correspondingly more complex equations of motion.

Writing the differential equation as the pair

$$\dot{x}_1 = x_2, \qquad \dot{x}_2 = u \tag{11.15}$$

the modified functional becomes

$$J^*[x_1, x_2, u, T] = T^2 - [\lambda x_1 + \mu x_2]_0^T + \int_0^T [(x_1{}^2 + u^2 + \lambda x_2 + \mu u) + \dot{\lambda} x_1 + \dot{\mu} x_2]\, dt,$$

with two Lagrange multipliers λ, μ and where an integration by parts of the last two terms has been performed. The Euler equations in terms of the Hamiltonian

$$H = x_1{}^2 + u^2 + \lambda x_2 + \mu u \tag{11.16}$$

are now

$$0 = H_u = 2u + \mu, \qquad H_{x_1} = 2x_1 = -\dot{\lambda}, \qquad H_{x_2} = \lambda = -\dot{\mu}, \tag{11.17}$$

which must be solved together with (11.15). The boundary conditions at $t = T$ need a little care and a return to first principles is again necessary. (A perfectly general result can be obtained with a little effort, see problem 6.) For conditions at T the only variables that need be considered are for those variables that are not fixed at T. Writing $T = \bar{T} + \varepsilon\tau$, $x_2 = X_2 + \delta\alpha(t)$, $u = U + \theta v(t)$, with ε, δ, θ as parameters and $\alpha(0) = 0$ to satisfy the given boundary conditions at $t = 0$, the change in J^* to first order becomes

$$\Delta J^* = (2\bar{T} - \dot{\lambda} x_1 - \dot{\mu} x_2)\varepsilon\tau\Big|_{t=\bar{T}} - (\mu\alpha)\,\delta\Big|_{t=\bar{T}} + (H + \dot{\lambda} x_1 + \dot{\mu} x_2)\varepsilon\tau\Big|_{t=\bar{T}} + \int_0^{\bar{T}} [H_u v\theta + (H_{x_2} + \dot{\mu})\alpha\delta]\, dt.$$

The integral term is zero since the Euler equations (11.17) are satisfied and the usual variational argument gives for the remaining terms at $t = \bar{T}$

$$\mu(\bar{T})\alpha(\bar{T}) = 0, \qquad 2\bar{T} + H(\bar{T}) = 0.$$

Since $\alpha(\bar{T})$ is unspecified, the natural boundary condition is $\mu(\bar{T}) = 0$, while the other equation

$$2\bar{T} + H(\bar{T}) = 0 \tag{11.18}$$

gives an equation for the final time $\bar{T}$.

The equations (11.15), (11.17) can be written conveniently in matrix form as

$$\frac{d}{dt}\begin{bmatrix} x_1 \\ x_2 \\ \mu \\ \lambda \end{bmatrix} = \begin{bmatrix} 0 & 1 & 0 & 0 \\ 0 & 0 & -\frac{1}{2} & 0 \\ 0 & 0 & 0 & -1 \\ -2 & 0 & 0 & 0 \end{bmatrix} \begin{bmatrix} x_1 \\ x_2 \\ \mu \\ \lambda \end{bmatrix}, \tag{11.19}$$

with the control variable given by $u = -\frac{1}{2}\mu$. This set of equations can be solved by writing all the variables as (ae^{kt}) and it will be found that k must take the values of the four roots of $k^4 = -1$. The four arbitrary constants are then computed from the conditions $x_1(0) = 1$, $x_2(0) = 0$, $x_1(T) = 0$, $\mu(T) = 0$ and the final time, T, is then obtained from (11.18).

It is clear that, even in this example with simple linear equations to solve, the solution is no trivial matter, although the method is well defined. It should be emphasized again that the major difficulty in a computational solution of (11.19), or corresponding equations in more difficult cases, is the two point nature of the boundary conditions. One possible method of solution is a 'shooting method'. In addition to the given conditions for x_1 and x_2 at $t = 0$, initial values are chosen for μ and λ, the solution is then possible by a forward integration method to obtain values of $x_1(T)$ and $\mu(T)$. These will not be zero, so the initial values of μ and λ are adjusted to try to bring them to zero. A Rosenbrock hill climbing technique is then frequently employed in this adjustment by minimizing the distance between the computed and actual values of $x_1(T)$ and $\mu(T)$. In this particular problem the solution is further complicated by also having T unspecified.

The most general problem considered in section 11.1 will not be looked at here but it is left as a rather difficult exercise in problem 8.

11.3* Constrained Control Parameters

To obtain about the simplest constrained problem, consider the first example in 11.2, of finding the minimum of

$$J[u] = \int_0^1 \varphi(x, u, t)\, dt$$

subject to

$$\dot{x} = f(x, u, t)$$

and boundary condition $x(0) = a$. Now add the constraint on the control variable u

$$g(x, u, t) \geq 0. \tag{11.20}$$

The analysis in section 11.2 can be followed through to equation (11.12) with the same definition of $H = \varphi + \lambda f$. Then, however, the Euler equations cannot be written down immediately since these equations demand that the variations in x and u in the modified functional J^* are independent. This is clearly *not* necessarily the case since (11.20) must be satisfied. Provided the values are well away from the boundaries of (11.20) there is no difficulty but when the boundary is encountered the variations are then dependent on each other. In particular it is usual to consider x as independent so that u must then be restricted. The Euler equation $\dot{\lambda} = -\partial H/\partial x$ follows and also the basic equation $\dot{x} = \partial H/\partial \lambda = f$ is valid. However, the other equation $\partial H/\partial u = 0$ does not necessarily follow. This corresponds to the fact that the local optimum may be outside the feasible region and hence the optimum must be on to the boundary of this region. The solution to this problem is a non-trivial exercise and takes three or four pages of detailed analysis in an advanced textbook such as Sage (1968). Consequently this solution will just be quoted and it will be left to the reader to consult further for the proofs.

The solution is called the *Pontryagin minimum* (maximum if $H = -\varphi + \lambda f$) *principle*. The optimum control variable $\bar{u}$ and the corresponding $\bar{x}$ satisfy

$$H(\bar{x}, \bar{u}, t) \leq H(x, u, t) \tag{11.21}$$

for all t and for all u that belong to the feasible region defined by (11.20). The Hamiltonian, H, must thus provide a global minimum in the feasible region. The other equations

$$\dot{\lambda} = -\frac{\partial H}{\partial x}, \qquad \dot{x} = \frac{\partial H}{\partial \lambda} = f$$

must be satisfied along the optimum path, at least where the derivatives are well defined. The natural boundary conditions and terminal equation for the final time T also hold as appropriate. This same minimum principle is equally valid for the general case when $\mathbf{x}$ and $\mathbf{u}$ have many components.

It is now possible to solve a constrained problem, at least in principle. To illustrate its use consider example 10 slightly rewritten from section 11.1 as: minimize the functional

$$I[u] = \int_0^T 1\,dt$$

subject to

$$\dot{x} = y, \qquad \dot{y} = u(t),$$

with boundary conditions

$$x(0) = a, \qquad y(0) = 0, \qquad x(T) = 0, \qquad y(T) = 0$$

and constraint

$$|u| \leq 1.$$

Now the Hamiltonian is just

$$H = 1 + \lambda y + \mu u$$

and this is to be minimized with respect to u. This is achieved by putting $u = -1$ if $\mu > 0$ and $u = +1$ if $\mu < 0$. Thus the control variable $u(t)$ depends only on the sign of μ and changes from one extreme value of the constraint to the other as μ changes sign. Such controls are usually referred to in the literature as *bang-bang* controls. The only crucial question is, *when* does the sign change?

The other equations of the system hold so that

$$\dot{\lambda} = -\frac{\partial H}{\partial x} = 0, \qquad \dot{\mu} = -\frac{\partial H}{\partial y} = \lambda,$$

which must be solved together with

$$\dot{x} = y, \qquad \dot{y} = u$$

and the given boundary conditions. Solving for the cases $u = \pm 1$ (the $\pm$ being followed through carefully) gives

$$\lambda = A, \qquad \mu = At + B \tag{11.22}$$

$$y = \pm t + C, \qquad x = \pm\tfrac{1}{2}t^2 + Ct + D, \tag{11.23}$$

with A, B, C, D arbitrary constants. Since the final time T is an unknown, eliminate t between the equations (11.23), to give

$$x - D = \pm\tfrac{1}{2}(y^2 - C^2). \tag{11.24}$$

These parabolas are plotted in figures 11.2(a), 11.2(b) in the (x, y) plane for the + and − cases respectively. The point defined by (11.23), or equivalently (11.24), must finally pass through the origin (0, 0) so that it must eventually follow the heavy line in either figure 11.2(a) or 11.2(b). Since $x(0) = a$, $y(0) = 0$ the initial point is at P shown in figure 11.2(b). This point then follows the curve shown for $u = -1$ until it reaches the heavy curve of

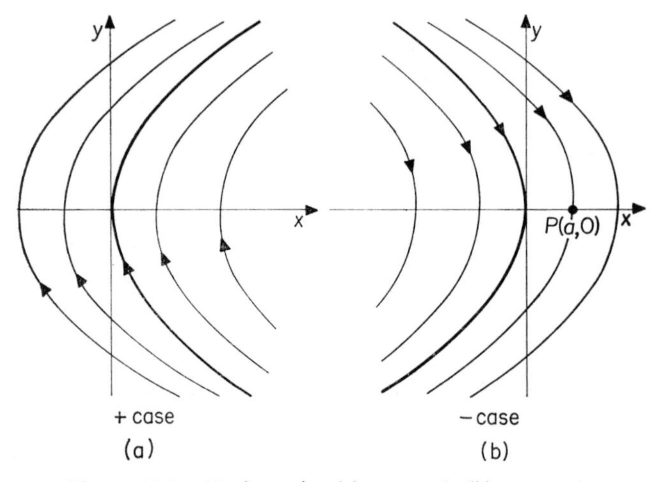

Figure 11.2 Trajectories (a) $u = +1$, (b) $u = -1$.

figure 11.2(a). The control variable then changes to $u = +1$ and the heavy line is followed into the origin. In figure 11.3 this piecewise differentiable function is drawn explicitly

Once this curve PSO has been established, all the other data for the problem can then be computed. It will be observed that a similar solution can be constructed for any starting conditions. The appropriate plus or minus curve is followed to one of the branches of the heavy or *switching* curve illustrated in figure 11.3, and then along this switching curve into the origin.

The typical feature of this problem is the fact that x, y and u have either discontinuities or discontinuous gradients at the point S. Because the equations (11.22), (11.23) are simple solutions it has been possible to avoid these discontinuities. However, a numerical procedure is normally required for the solution of the basic equations and it must be designed to cope with such discontinuities.

The problem just studied is about as difficult as can be dealt with, without recourse to a great deal of new mathematics or new computational procedures. Suffice it to add that hill climbing techniques, finite difference schemes, quasilinearization and dynamic programming have been used in various combinations to attempt a detailed solution of such problems. Further work and practical examples may interest the reader and may be found in the problems and such textbooks as Sage (1968) or Noton (1965).

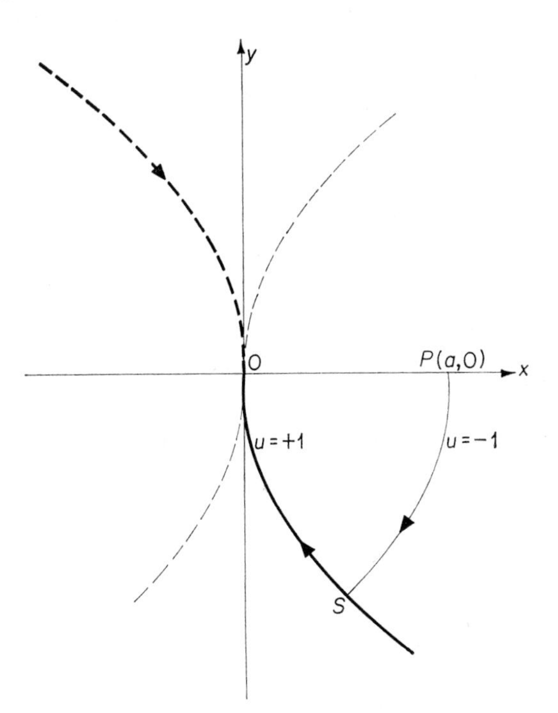

Figure 11.3 Final solution *PSO*.

PROBLEMS

1. If the $H(x, u, t)$ defined in (11.12) is explicitly independent of t, i.e. $H = H(x, u)$, show that the equations (11.13), (11.14) imply that $\partial H/dt = 0$. Show that this result holds for the corresponding vector case $H(\mathbf{x}, \mathbf{u})$. (Note that this implies that H is constant along the optimal trajectory.)
2. Find the function $u(t)$ which minimizes

$$J[u] = \int_0^1 u^2\, dt,$$

where

$$\dot{x} = u + ax, \qquad x(0) = 1,$$

and a is a constant.

3. Solve equations (11.19) as suggested, that is by making all the variables proportional to exp (kt). Try to fit the appropriate boundary conditions.
4. Use the method of section 11.2 to solve the problem defined by (11.1) and (11.2). Compare with the dynamic programming solution in section 10.4.
5. The cost function

$$J[u] = \tfrac{1}{2}[x(T)]^2 + \tfrac{1}{2}\int_0^T u^2(t)\, dt$$

is to be minimized subject to

$$\dot{x} = -x + u \quad \text{and} \quad x(0) = x_0$$

and T is fixed. Show that this is equivalent to minimizing

$$J^*[u] = \int_0^T (x\dot{x} + \tfrac{1}{2}u^2)\, dt,$$

subject to the constraint.
Hence deduce that the optimal trajectory is

$$x(t) = x_0 e^{-t} + \lambda_0 \sinh t,$$

where λ_0 is a constant.
6. Show that the terminal conditions for the general problem of optimizing

$$J[u] = \theta[x(T), T] + \int_0^T \varphi(x, u, t)\, dt,$$

subject to

$$\dot{x} = f(x, u, t), \qquad x(0) = a$$

and a terminal manifold $N(x(T), T) = 0$, are

$$\left(\frac{\partial \theta}{\partial x} - \lambda + \mu \frac{\partial N}{\partial x}\right)_{t=T} = 0, \qquad \left(\frac{\partial \theta}{\partial t} + H + \mu \frac{\partial N}{\partial t}\right)_{t=T} = 0,$$

where μ is further Lagrange multiplier.
7. Obtain the minimum of

$$J[u] = \int_0^1 u^2\, dt,$$

where

$$\dot{x} = y, \qquad \dot{y} = u, \qquad x(0) = y(0) = 0$$

and $x^2(1) + y^2(1) = 1$.

8. For the general problem of the type considered in section 11.2 find the conditions for optimizing

$$J[\mathbf{u}] = \theta[\mathbf{x}(t), t]\Big|_{t_0}^{t_f} + \int_{t_0}^{t_f} \varphi(\mathbf{x}, \mathbf{u}, t)\, dt,$$

with

$$\dot{\mathbf{x}} = \mathbf{f}(\mathbf{x}, \mathbf{u}, t)$$

and initial and terminal manifolds

$$M[\mathbf{x}(t_0), t_0] = 0, \qquad N[\mathbf{x}(t_f), t_f] = 0.$$

Show that the equations for the optimal trajectory take the form (using a generalized notation),

$$\dot{\boldsymbol{\lambda}} = -\frac{\partial H}{\partial \mathbf{x}}, \qquad \dot{\mathbf{x}} = \frac{\partial H}{\partial \boldsymbol{\lambda}}, \qquad \frac{\partial H}{\partial \mathbf{u}} = 0,$$

where $H = \varphi + \boldsymbol{\lambda}^T\mathbf{f}$ and $\boldsymbol{\lambda}^T = (\lambda_1, \ldots, \lambda_n)$ is a set of Lagrange multipliers. Show also that the initial and terminal conditions are analogous to those in problem 6.

9. A rocket ascends vertically under a constant gravitational attraction and with negligible aerodynamic forces. The thrust acts vertically downwards and the equations of motion become

$$\dot{h} = v, \qquad \dot{v} = -g + \frac{c\beta}{m}, \qquad \dot{m} = -\beta,$$

where h is the height, v the vertical velocity, m the rocket mass and c is a constant. The β is the propellant mass flow and is subject to the constraint $0 \le \beta \le \beta_{max}$. At $t = 0$ the mass, height and velocity are known. Show that to maximize the height, the control $\beta(t)$ is of bang-bang type. Deduce a form for the switching curve.

Chapter XII

Related Minimization Methods

12.1 Least Squares

This is perhaps the most obvious of the minimization methods which can be used to solve the differential equation

$$Lu = 0 \quad \text{in } R. \tag{12.1}$$

An approximate solution u_1 is chosen to satisfy the boundary conditions but not necessarily the equation, so that

$$Lu_1 = r_1.$$

The method then minimizes the square of the residual

$$\int_R r_1{}^2 \, dV,$$

where the integration is over the whole of R. Since an exact solution of (12.1) would make $r_1 = 0$ in R, the minimization can be expected to give a reasonable approximation.

Example 69 Using the least squares method solve

$$u'' = u^3$$

with $u(0) = 1$ and $u(1) = 0$.

The function

$$u = (1 - x)(1 + ax + bx^2 + \cdots)$$

satisfies the boundary conditions. Take the required approximation as the simplest one, that is, just involving one unknown parameter a. Now

$$u'' - u^3 = -2a - (1 - x)^3(1 + ax)^3,$$

so that it is required to minimize

$$f(a) = \int_0^1 [2a + (1 - x)^3(1 + ax)^3]^2 \, dx$$

with respect to a. Performing the integrations, this minimization requires the solution of the quintic equation

$$f'(a) = 0 = 1.107 + 9.319a + 0.671\,4a^2 + 0.140\,3a^3 + 0.005\,411a^4 + 0.000\,499\,5a^5.$$

The appropriate root of this equation is $a = -0.119\,8$ so the best least squares quadratic approximation is

$$u = (1 - x)(1 - 0.119\,8x).$$

Two points should be noted from this example, firstly the method is quite capable of dealing with non-linear differential equations and secondly it is not necessary to compute a functional giving the differential equation as its Euler equation. Both these points are extremely valuable in tackling new problems and they should be noted carefully.

An alternative use of the least squares idea can be seen in the next example where the differential equation is satisfied exactly but the boundary conditions are not.

Example 70 Solve $\nabla^2 u = 0$ in the region $\{R: |x| \leq 1, |y| \leq 1\}$ with the conditions

$$u = y \text{ on } x = \pm 1; \qquad u = 1 \text{ on } y = 1; \qquad u = -x^2 \text{ on } y = -1.$$

Now it is well known that the functions x, y, $x^2 - y^2$, xy, $x^3 - 3xy^2$, $3x^2y - y^3, \ldots$ satisfy the Laplace equation. It is clear from the boundary conditions that the solution is an even function of x and hence involves x^2 only. Thus try as an approximate solution

$$u = ay + b(x^2 - y^2). \tag{12.2}$$

The Laplace equation is satisfied exactly, so now try to satisfy the boundary conditions as accurately as possible by minimizing

$$\int_B (u_{\text{exact}} - u_{\text{approx}})^2 \, ds,$$

where the integral is taken round the boundary B of R. This integral in the present case is

$$f(a, b) = \int_{-1}^{+1} [-x^2 + a - b(x^2 - 1)]^2 \, dx + 2\int_{-1}^{+1} [y - ay - b(1 - y^2)]^2 \, dy + \int_{-1}^{+1} [1 - a - b(x^2 - 1)]^2 \, dx.$$

Performing all the integrations, $f(a, b)$ can be computed and the minimization gives

$$\frac{\partial f}{\partial a} = 0 = \frac{8}{3}(-3 + 4a)$$

$$\frac{\partial f}{\partial b} = 0 = \frac{32}{15}(1 + 4b).$$

Thus the best approximation that can be achieved using this method on (12.2) is

$$u = \tfrac{3}{4}y - \tfrac{1}{4}(x^2 - y^2).$$

There are many variations of this method that can be used as an alternative to more traditional approaches. One of the main advantages of this method, particularly as applied in example 70, is that a close check on accuracy can be maintained since such simple functions are used.

12.2 Trefftz Method

A method similar to the above is due to Trefftz. Again the differential equation is satisfied exactly while the boundary conditions are approximated by adjusting free parameters. The method does require that an appropriate functional is available giving the required equation as its Euler equation; this is a disadvantage compared with the least squares method. However, for a positive definite operator, requiring the minimum of a functional, the Trefftz method gives a lower bound on this minimum while the Ritz method gives an upper bound. This information can be very useful, particularly in eigenvalue problems (cf. section 8.1.3).

Consider the straightforward problem of solving the Poisson equation

$$\nabla^2 u = -\rho \quad \text{in } R \tag{12.3}$$

with the Dirichlet conditions

$$u = g \quad \text{on } B,$$

where B is the boundary of the closed region R and g is known function. It has been shown that the functional

$$I[u] = \int_R [(\nabla u)^2 - 2\rho u]\, dV \tag{12.4}$$

has (12.3) as its Euler equation. Choose functions $\bar{u}, u_1, u_2, \ldots, u_N$ which satisfy

$$\nabla^2 \bar{u} = -\rho \quad \text{and} \quad \nabla^2 u_i = 0, \qquad i = 1, 2, \ldots, N,$$

and let U be the exact solution of (12.3). Now the function

$$\theta = -U + \bar{u} + \sum_{i=1}^{N} a_i u_i \tag{12.5}$$

will be zero if the exact solution can be written in terms of $\bar{u}$ and u_i. Thus it is required to adjust the constants a_i to make θ as small as possible in the sense of minimizing (12.4), which is taken as a measure of the distance of θ from the zero function. But $I[\theta]$ is just a function of the a_i's so that

$$0 = \frac{\partial I[\theta]}{\partial a_j} = \int_R (2 \nabla u_j \, . \, \nabla\theta - 2\rho u_j) \, dV$$

or

$$0 = \int_R [\theta \nabla^2 u_j - \nabla \, . \, (\theta \nabla u_j) - \rho u_j] \, dV.$$

The function u_j satisfies the Laplace equation, however, and using the divergence theorem on the second term gives

$$0 = \int_B \left(-U + \bar{u} + \sum a_i u_i\right) \nabla u_j \, . \, d\mathbf{S} - \int_R \rho u_j \, dV.$$

The final set of equations can then be obtained by using the fact that $U = g$ on B, as

$$\sum_{i=1}^{N} a_i \left(\int_B u_i \nabla u_j \, . \, d\mathbf{S}\right) = \int_B (g - \bar{u}) \nabla u_j \, . \, d\mathbf{S} + \int_R \rho u_j \, dV, \quad j = 1, 2, \ldots, N. \tag{12.6}$$

Example 71 Repeat example 70 using the Trefftz method.

In the present notation $\rho = 0$ and $\bar{u} = 0$ and the same approximate function (12.2)

$$u = a_1 y + a_2(x^2 - y^2)$$

will be used. Evaluating the integrals in (12.6) gives the equations

$$0a_1 + 8a_2 = -4$$

$$8a_1 + 0a_2 = \frac{16}{3}$$

so that

$$u = \tfrac{2}{3}y - \tfrac{1}{2}(x^2 - y^2).$$

This result differs from the least squares result but it gives the 'best' that can be obtained from the Trefftz method. The reader should compare these functions round the boundary B from the two approaches to check how accurately the exact boundary conditions are approximated.

Although only a simple illustration has been used, the basic method can be followed through with whole classes of problems to give equations similar to (12.6). This method is particularly useful if the boundaries or the boundary conditions are awkward.

12.3 Galerkin's Method

As stressed above, it is particularly important to find methods which avoid constructing a functional but still use the same basic approximation procedures. B. G. Galerkin produced a particularly powerful approach which includes the Rayleigh–Ritz method as a special case. It fits into a class of methods which is commonly called the method of weighted residuals and is reviewed by Finlayson and Scriven (1966). The generality of these methods has meant that, except in well established fields, they have superseded Rayleigh–Ritz techniques. In particular, the method has been used increasingly for non-linear problems, where it provides a very convenient framework for a numerical solution.

Consider the differential equation

$$L\omega = 0 \quad \text{in } R, \tag{12.7}$$

with $\omega = g$ on the boundary B of R and g is a given function. Find a complete set of functions $\{u_i\}$ with $u_0 = g$ on B and $u_i = 0$ $(i = 1, 2, \ldots)$ on B, then it is required to find an approximate solution

$$\omega_N = u_0 + \sum_{i=1}^{N} a_i u_i. \tag{12.8}$$

Substituting (12.8) into (12.7)

$$L\left(u_0 + \sum a_i u_i\right) = r_N,$$

where r_N can be considered as the residual, which should be zero for an exact solution. But it is known that the only function φ for which,

$$\int_R \varphi u_i \, dV = 0, \quad \text{all } i,$$

with respect to a complete set, is the zero function.

Thus the residuals r_N are made to satisfy the conditions (12.9) for $u_1, u_2, \ldots, u_N$,

$$\int_R u_j L\left(u_0 + \sum_{i=1}^{N} a_i u_i\right) dV = 0, \qquad j = 1, 2, \ldots, N, \tag{12.9}$$

which gives N equations for the N unknowns $a_1, \ldots, a_N$.

Example 72 Use Galerkin's method to find an approximate solution of

$$u'' = u^3$$

with $u(0) = 1$, $u(1) = 0$. (cf. example 69).

The same functions used in example 69 will be considered so that

$$\omega_1 = (1 - x) + ax(1 - x)$$

and giving

$$r_1 = \omega''_1 - \omega_1{}^3 = -2a - (1 - x)^3(1 + ax)^3.$$

The conditions (12.9) now read for just the one constant a,

$$\int_0^1 [-2a - (1 - x)^3(1 + ax)^3]x(1 - x)\, dx = 0,$$

or evaluating the integrals

$$4a^3 + 27a^2 + 912a + 84 = 0.$$

The only real root of this cubic is $a = -0.092\,35$ giving an approximate solution

$$\omega = (1 - x)(1 - 0.092\,35x),$$

which may be compared with the least squares result in example 69. A comparison should be made of the difficulties encountered in the two approaches and it will be found that the Galerkin method requires the least computational work; this trend applies in almost all cases.

For a linear, self-adjoint, positive definite operator L the Ritz and Galerkin method become identical. Suppose it is required to solve

$$Lu = f \quad \text{in } R, \tag{12.10}$$

with $u = 0$ on the boundary B of R. It is always possible to reduce a Dirichlet problem to one of zero boundary conditions so this is no restriction. Now it may be shown (cf. section 8.1) that this equation can be deduced as the Euler equation of

$$I[u] = \int_R (uLu - 2uf)\, dV. \tag{12.11}$$

The Rayleigh–Ritz method defines a complete set of functions $\{u_i\}$ satisfying zero boundary conditions and then tries to compute an approximate solution of the form

$$u = \sum_{i=1}^{N} a_i u_i, \tag{12.12}$$

with the a_i computed from the minimum of (12.11). Defining

$$L_{ij} = \int_R u_i L u_j \, dV \quad \text{and} \quad f_i = \int_R f u_i \, dV$$

(12.11) becomes

$$I[u] = \sum_{i,j} L_{ij} a_i a_j - 2 \sum_i a_i f_i$$

and evaluating $\partial I / \partial a_i = 0$ gives the set of equations

$$\sum_{j=1}^{N} L_{ij} a_j = f_i, \qquad i = 1, 2, \ldots, N. \tag{12.13}$$

The Galerkin method for this problem gives precisely the same result since substituting (12.12) into (12.10)

$$r_N = \sum_{j=1}^{N} a_j L u_j - f,$$

the Galerkin conditions are

$$\int_R \left(\sum a_j L u_j - f \right) u_i \, dV = 0, \qquad i = 1, \ldots, N$$

or using the above notation

$$\sum_{j=1}^{N} L_{ij} a_j = f_i, \qquad i = 1, \ldots, N.$$

Since these equations are identical with (12.13), the Galerkin method includes the Ritz method, at least in respect of equation (12.10). It should be noted that the linearity and self-adjointness of L are used continually in the above analysis and is crucial for the equivalence. The Galerkin method is clearly more general, since the same approach can be used for a quite general operator L, although it should be noted that the convergence of the method cannot be assured in the more complex cases. Despite this, the method is found to work very well and is perhaps the most important of this type of approximation technique.

12.4* Partial Integration

A very standard procedure in differential equations is to reduce a partial differential equation to the solution of more amenable ordinary differential equations. This can be achieved by assuming a solution of the form $u = \varphi(x, y)f(x)$ (for a two variable problem) where φ is a known function, while f is unknown and to be determined. If good physical or intuitive reasons can be found to suggest the form of φ then the method can be very accurate. This assumed form is then substituted into the appropriate functional, the integrations carried out with respect to y variable and the remaining functional involves the single unknown function f. In some cases the Euler equation for f can be solved explicitly, while in other cases the method just gives a useful idea on how to proceed further.

For the Poisson equation in a rectangle a solution of the form $u = \Sigma f_n(x) \sin (ny/a)$ can be assumed and very fast Fourier analysis routines can be used to make a solution by this method into a viable, very competitive numerical procedure. Several more useful examples can be found in Schechter (1967) or Kantorovich and Krylov (1958). The following example, which illustrates both the method and its difficulties, is to find the eigenvalues of the wave equation,

$$\nabla^2 \psi = \frac{1}{c^2} \frac{\partial^2 \psi}{\partial t^2}$$

in the triangular region, R, bounded by the lines $y = 0$, $x = 1$, $y = x$. These eigenvalues just give the frequencies, or notes, obtained from a triangular drum. They are obtained by looking for periodic solutions

$$\psi = u(x, y) \sin \mu t$$

so that

$$\nabla^2 u + \lambda^2 u = 0, \tag{12.14}$$

where $\lambda^2 = \mu^2/c^2$. This Helmholtz equation has to be solved with the boundary condition $u = 0$ on the whole of the boundary. It was shown in section 8.1.3 that the Euler equation of the functional

$$I[u] = \int_R [(\nabla u)^2 - \lambda^2 u^2] \, dx \, dy \tag{12.15}$$

is just (12.14). A suitable function, which satisfies the condition that $u = 0$ on $y = 0$ and $y = x$, may be taken as

$$u = y(y - x)f(x), \tag{12.16}$$

where $f(x)$ must satisfy $f(1) = 0$ and $f(0)$ bounded. Putting this into (12.15) and integrating with respect to y from $y = 0$ to $y = x$ gives after a little manipulation

$$I[y(y - x)f] = J[f] = \frac{1}{30}\int_0^1 [x^5 f'^2 + 5x^4 ff' + f^2(20x^3 - \lambda^2 x^5)]\, dx.$$

Now the problem is to find and solve the Euler equation for f obtained from this functional. Using Result IV gives

$$0 = \frac{x^3}{15}[x^2 f'' + 5xf' + (-10 + \lambda^2 x^2)f]. \tag{12.17}$$

The optimum solution of the type (12.16) now requires the solution of (12.17). A power series solution can be tried but these are usually hard work; a numerical procedure is possible but difficult because of the singularity at $x = 0$; a variational solution of (12.17) implies that such a method could have been used at the starting point and the above analysis is redundant. The differential equation can be solved exactly in fact since it is reminiscent of the Bessel equation, so a reduction can be tried of the form $f = x^n g$. Putting $n = -2$, $f = x^{-2}g$, then (12.17) becomes

$$x^2 g'' + xg' + (-14 + \lambda^2 x^2)g = 0,$$

with solutions $J_{(14)^{\frac{1}{2}}}(\lambda x)$, $J_{-(14)^{\frac{1}{2}}}(\lambda x)$. Now the solution must be bounded at the origin so only the first of these is valid since then $g \sim x^{(14)^{\frac{1}{2}}}$ and hence $f \sim x^{(14^{\frac{1}{2}} - 2)}$. The eigenvalues or frequencies are now determined from the other boundary condition that $f = 0$ at $x = 1$. Thus the solutions of $J_{(14)^{\frac{1}{2}}}(\lambda) = 0$ are required. This equation will still require a fair bit of numerical computation to extract the roots but at least the Bessel function has extensively quoted properties and a mass of tabulated values are available. The whole of this information can now be used to understand the approximate solution of form (12.16).

Problems

1. Use (i) the least squares method, (ii) the Galerkin method to compute approximate solutions of

$$y'' + y + \cos x = 0, \qquad y(-1) = y(1) = 0.$$

Compare with the exact solution

$$y = \tfrac{1}{2}\tan 1 \cos x - \tfrac{1}{2}x \sin x.$$

2. For the differential equations

(i) $uu'' + 2u'^2 + u^2 = 0, \qquad u(0) = 0, u(1) = 1$

(ii) $u'' + (1 + 0.1u'^2)u = 1, \qquad u(0) = 0, u(\tfrac{1}{2}\pi) = 1,$

compare the results obtained from the simplest approximations for the least squares and Galerkin methods.

3. For the differential equation in section 8.6

$$u'' = \sinh u, \qquad u(0) = u_0, \qquad u \to 0 \text{ as } x \to \infty$$

compare the values obtained for α in the approximate solution of the form $u_0 \exp(-\alpha x)$ when (i) the least squares, (ii) Galerkin methods are used.

4. In the pipe flow problem of section 9.1.1 it was required to solve

$$\nabla^2 v = -\frac{P}{\mu} = \text{constant}$$

in a region $\{R: |x| \leq a, |y| \leq a\}$ and with $v = 0$ on the boundary of R. Find a sequence of functions that satisfy the equation exactly and then use the method of least squares round the boundary and also the Trefftz method to find approximations to the solution. Compare with the quoted values in the text of section 9.1.1.

5. In plane polar coordinates (r, θ) a circular plate occupies the region $|r| \leq a$. The temperature of the plate $T(r, \theta)$ satisfies the steady state equation

$$r\frac{\partial}{\partial r}\left(r\frac{\partial T}{\partial r}\right) + \frac{\partial^2 T}{\partial \theta^2} = 0.$$

The upper edge of the plate $0 \leq \theta \leq \pi$, $r = a$ is held at $T = T_0$ while the lower edge $\pi < \theta < 2\pi$, $r = a$ is held at $T = 0$. Use the Trefftz and least squares method to approximate to the solution.

6. It is required to solve the eigenvalue problem

$$xy'' + y' + \lambda^2 \frac{x}{a^2}\left(1 - \frac{x^2}{a^2}\right) y = 0$$

with $y(a) = 0$, $y(0)$ bounded. Adapt the least squares and Galerkin method to estimate the lowest eigenvalue λ. The accurate value is $\lambda = 2.704$, while the Rayleigh–Ritz method gives 2.76 with the function $A \cos(\pi x/2a)$. (Note that this problem occurs when solving for the temperature in the entrance region of a pipe.)

7. Use the Galerkin method to find an approximate solution to the integral equation

$$1 = \int_0^x \frac{u(t)\,dt}{(x^2 - t^2)^{\frac{1}{2}}}, \qquad 0 \leq x \leq 1.$$

8. Look for solutions of the Helmholtz equation

$$\nabla^2 u + \lambda^2 u = 0 \quad \text{in } |r| \leq a,$$

with $u = 0$ on $r = a$, where (r, θ) are plane polar coordinates. Use the partial integration method of section 12.4 to find the 'best' solution of the form (i) $f(r) \sin \theta$, (ii) $(a^2 - r^2)g(\theta)$.

Chapter XIII*

Problems for Further Study

This final chapter contains only the statement of problems for further study, with little attempt at any analysis. Some of these problems have already been touched on in the text. For instance in section 7.5 the maximum enclosed area problem was studied in some detail and the initial comments in that section are relevant to sufficiency. It was noted that the Euler equation only gives a necessary condition and it is not sufficient. That is, if a functional is known to have an extreme function, then this function must satisfy the Euler equation; this is the necessity. If, however, a function satisfies the Euler equation then this is no guarantee that the function is an extremum; it is only when the sufficiency has been proved that this guarantee can be given. The proof of sufficiency is usually an extremely difficult piece of work and normally limited to a single problem or, at best, a narrow class of problems. Some of the simpler available results can be found in Gelfand and Fomin (1963). For practical purposes, in any but well worn situations, this analysis is out of the question and physical or geometric intuition has to be relied on to support the solution obtained. This is usually the best that can be done but it is, of course, no substitute for the proof of a basic sufficiency theorem.

In the direct methods, such as Rayleigh–Ritz or Galerkin, the difficulty of these methods is whether the series of approximations converges or not. In section 5.2.1 comments were made on the general problem of the convergence of sequences of functions. Example 31 provides a typical result in which a sequence of functions, each infinitely many times differentiable, tends to a function which is not differentiable at one particular point. Readers familiar with Fourier series will have appreciated already that this sort of behaviour can occur. Given this sort of difficulty the problem is, how can the theory be developed to cope with this situation? There appear to be two main ways; either only accept solutions which are very well behaved or extend the class of functions studied and accept discontinuous functions. In the first case the type of solution required must be stated quite clearly. If this course of action is taken then it often happens that no solution of the required type exists. If, however, the second development is favoured there is much more likelihood of a solution being found, but serious problems of definition and computation arise. If a discontinuity in the solution occurs at a single point, or several discrete points, the situation can be treated fairly straightforwardly; the Weierstrass–Erdmann conditions have been established for just such a situation (see Akhiezer (1962)). It is clear that some special conditions are required at these points since the Euler equation cannot be defined there. On the other hand the direct methods can lead to much more complex situations than this. The class of 'functions' defined by, say, all twice differentiable functions and all limits of sequences of such functions widens the problem considerably. (The reader may compare the situation of rational numbers and sequences of rationals which leads to defining the much larger class of irrationals.) This wider class is usually called the class of generalized functions and a new approach must be made to the differentiation and integration of such functions. The integration method that must be considered is the Lebesgue integral and the Frechet derivative gives a method of overcoming some of the differentiation difficulties. There are considerable advantages in using these ideas since it enables the whole of ordinary differentiation and functional differentiation, including the whole of the calculus of variations, to be embraced in identical formulae.

Computationally it is difficult to see how these abstract ideas can be useful directly, since almost all numerical techniques rely on local approximation by polynomials and the Weierstrass approximation theorem no longer holds if functions are not continuous. For these reasons, from a practical viewpoint, these ideas on convergence, generalized functions etc, are considered by some workers to be somewhat esoteric. The sequences they consider either happen to behave themselves beautifully or they can only conceive of finding one or two terms in the sequence and hence they are prepared to make do with these. However, in subjects such as control theory, discontinuous solutions are obtained in practical cases, see section 11.3, so it is absolutely essential to understand such situations and to be able to work with them with confidence. Therefore, once an initial understanding has been established in the basic problems of optimization theory and the mechanics of the various methods, it is essential to look further into the problems discussed above. This leads to a more abstract approach to the subject since all the definitions used must be much more carefully specified. These ideas give a deeper understanding of the basic methods of optimization and incidentally prove to be useful in other fields. As a suitable text for a start on this work the reader is referred to Luenberger (1969).

Perhaps the fastest moving of all the fields studied in this book is in the hill climbing methods and computer aspects of the various other methods. This is basically because these methods are continually trying to keep pace with the extremely rapid development of electronic computers. Increase in speed and storage has made new methods viable where in the past they could not even be considered. The need to use this increased power as efficiently as possible also leads to the continual modification of existing methods and programmes. Indeed it often appears that with each new issue of the relevant journals an old method has been superseded and a new variant has appeared. While this development is very healthy, it is hard to keep track of all these methods and to write or obtain a computer implementation of each of them is an impossibility. Over a longer period, however, one or two of these new methods establish themselves as the most robust and versatile and are then retained as standard techniques. It is usually better for the more inexperienced worker in the field to wait for a review book, such as Lootsma (1972), since these articles contain the distilled wisdom of one or more very experienced workers.

In many realistic problems a basic difficulty is that the information which is supposed known is not deterministic. For instance a production cost may be known as a statistical distribution or a machining operation may be subject to random fluctuations about a known mean. From such data it is not possible to get a deterministic maximum but only that such a maximum has a statistical distribution obtained from the calculations performed. These ideas have considerable importance in many economic and control theory applications and it is necessary to master the problem involved and the available techniques. These problems are less well documented but a suitable text to start such a study is Wilde (1964) or Hastings (1973).

These comments give just a few directions which may be followed. There are many unsolved problems along the way and new ideas are required both to solve them and to develop novel aspects of the work. For the active mathematical brain the subject of optimization can give considerable stimulus and satisfaction. For the worker interested in other fields it can provide an extremely useful additional tool in his armoury of mathematical weapons.

Answers and Comments on Problems

Chapter 1

1. Local minimum at $x = \frac{1}{3}$, local maximum at $x = 3$: global minimum at $x = \infty$, $y = 0$, global maximum at $x = 3$, $y = \frac{1}{2}$.
2. $(a^{2/3} + b^{2/3})^{3/2}$.
3. Point of inflexion on a rising curve.
4. Minimum at (0, 0).
6. Additional to Result II, the Jacobian matrix must be positive definite at the extremum, i.e. $\mathbf{y}^T\mathbf{J}\mathbf{y} > 0$ for all column vectors $\mathbf{y}$.
7. Local maximum at $(-2^{-\frac{1}{2}}, -2^{-\frac{1}{2}}, 0)$.

Chapter 2

1. Maximum at $x \approx 2.18$, $F = 0.026\,71$.
2. (i) Misses maximum.
   (ii) (0.1, 0.7) gives bracket.
12. $(1, 0, 0)$, $2^{-\frac{1}{2}}(0, -1, -1)$, $2^{-\frac{1}{2}}(0, -1, 1)$.
15. $(0, 0)$, $F = 1$; $(0, 0.5)$, $F = 0.5$; $(0.5, 0.5)$, $F = 0.25$; ....
    Booth's modification gives $(0, 0.45)$, $F = 0.505$; $(0.65, 0.61)$, $F = 0.154$; ....
16. Yes, No, Yes.

Chapter 3

1. Minima at (α, α), $(\alpha, -\alpha)$, $(-\alpha, \alpha)$, $(-\alpha, -\alpha)$ with $\alpha = 2^{-3/2}$.
2. No local minimum. Global minima at $x = \pm 1$.
3. Maximum at $(1, 0)$, $f = 2$.
7. $x = 30/7$, $y = 0$, $z = 20/7$, $f = 270/7$.
8. (i) 26
   (ii) $5\frac{1}{3}$ units of blue wool.

Chapter 4

2. The first principles deduction of the geodesic is not at all obvious.
3. Try $y = ax^2(x^3 - 3l^2x + 2l^3)$ and show $a = k/(120EI)$.

Chapter 5

1. $y = \sin x + a \sin 2x$ gives $a = 0$ (the exact result); $y = (2x/\pi) + \alpha x(2x/\pi - 1)$ gives $\alpha = -5\pi(40 - \pi^2)$.
3. $y = x - (11/9)(x - 1)(x - 2)$.
4. $u = \frac{1}{2}a^2(1 - Y^2)[1 + X - \frac{5}{8}(1 - x^2)]$ with $X = x/a$, $Y = y/a$.
5. $y = x^3 + 1.245x^3(1 - x)$.
6. $r = a(1 - \alpha t^2)$, $\theta = (h/a)(t + \beta t^2)$ gives a simple approximation but a *very complex* $f(\alpha, \beta)$ to minimize.
8. Because of the orthogonality of the u_p the L-matrix becomes diagonal and $a_p = -1/2\pi p(p^2 - \frac{1}{4})$.

Chapter 6

1. $y'' + y = 0$.
2. (i) $xy'' + y' + xy' = 0$
   (ii) $y^2(1 + y'^2) = K$ from (6.1)
   (iii) $r^2\dot{\theta} = C$, $\ddot{r} - r\dot{\theta}^2 + \mu/r^2 = 0$.
3. $\theta - \pi = \int_{-a}^{r} \left(\frac{C - t^2 - t^4}{1 + t^2}\right)^{\frac{1}{2}} dt$, $C =$ constant.
4. $h = H \log\left[\cos\left(\frac{a - 2x}{2H}\right)\Big/\cos\left(\frac{a}{2H}\right)\right]$.
5. $\sin^2\theta\,\varphi' = C(1 + \sin^2\theta\,\varphi'^2)^{\frac{1}{2}}$ with $\varphi = 0$ at $\theta = \frac{1}{2}\pi$ and $\varphi = 0$ at $\theta = \theta_0$. Solution $\varphi \equiv 0$.
6. (i) $0 = \operatorname{div}\{(\nabla u)/[1 + (\nabla u)^2]^{\frac{1}{2}}\}$.
   (ii) $\nabla^2 u = \frac{1}{2}g$.
7. $U = (2)^{\frac{1}{2}} \sin \pi x$, $J[U] = \pi^2$.
8. $\lambda = 15/a^2$.
9. Geodesic on a sphere; great circle.
10. $y = \cosh(1 - t)/\cosh 1$, $u = -\sinh(1 - t)/\cosh 1$. Boundary condition on λ is $\lambda(1) = 0$.
11. $\frac{d^2}{dx^2}\left(\frac{\partial f}{\partial u''}\right) - \frac{d}{dx}\left(\frac{\partial f}{\partial u'}\right) + \frac{\partial f}{\partial u} = 0.$
12. $u = 1 + ax + bx^2$, gives $a = -1\,536/308$, $b = 50/77$ and $u'(1) - u(1) = -27/77$.
13. $\partial u/\partial r = 0$.

Chapter 7

1. (i) Cycloid
   (ii) $y'(a) = 0$.

4. If $b = AC$ then $BC = a\left(1 - \dfrac{mga}{2\lambda b}\right)^{-1}$.

5. Stable if $(a_{11} + a_{22}) < 0$ and $(a_{11}a_{22} - a_{12}a_{21}) > 0$.
7. If $(kT/qJ) > 1$ only one solution, $R = 0$; if $(kT/qJ) < 1$, three solutions $0, \pm C$, the non-zero values giving the lowest free energy.
10. (i) The only non-zero symbols are

$$\begin{Bmatrix} & 1 & \\ 2 & & 2 \end{Bmatrix} = -\sin\theta\cos\theta, \qquad \begin{Bmatrix} & 2 & \\ 1 & & 2 \end{Bmatrix} = \cot\theta,$$

equations are

$$0 = \theta'' - \sin\theta\cos\theta\,\varphi'^2, \qquad 0 = \varphi'' + \cot\theta\,\theta'\varphi'.$$

(ii) The only non-zero symbols are

$$\begin{Bmatrix} & 1 & \\ 2 & & 2 \end{Bmatrix} = -r\sin^2\alpha, \qquad \begin{Bmatrix} & 2 & \\ 1 & & 2 \end{Bmatrix} = \frac{1}{r}$$

and the equations are

$$r'' - r\sin^2\alpha\,\theta'^2 = 0, \qquad \theta'' + \frac{1}{r}\theta' r' = 0.$$

11. $Ky = \cosh(Kx + C)$. Boundary conditions cannot be fitted for $\varepsilon = 0$. The solution in this case is discontinuous.

Chapter 8

1. (i) $x\dfrac{d^2}{dx^2} + 3\dfrac{d}{dx} + x$ (ii) $x\dfrac{\partial^2}{\partial x^2} + y\dfrac{\partial^2}{\partial y^2} + 2\dfrac{\partial}{\partial x} + 2\dfrac{\partial}{\partial y}$

   (iii) $\nabla^2 + k^2$.

2. $g = \dfrac{1}{a}\exp\left(\int \dfrac{b}{a}\,dx\right)$

3. $\int(\omega\nabla^2 u - \frac{1}{2}\omega^2)\,dV$. Note the self-adjoint boundary conditions required.
4. 15.00 compared with accurate value of 14.68.
6. (i) $\varphi_1 = \sin x$, $\varphi_2 = \sin 2x$ give the exact results 1 and 4
   (ii) $\varphi_1 = (1 - x)$, $\varphi_2 = x(1 - x)$ give $\lambda_1 = 5.5$, $\lambda_2 = 17.5$.

10. $\int_R (\nabla u)^2 \, dA - \int_{-a}^{+a} u^2(x, a) \, dx.$

12. Note that this gives a sufficiency proof for this functional.
14. $u_1 = 73.5$, $u_2 = 23.7$, $u_3 = 67.5$.

Chapter 9

1. $A = 0.292(P/\mu a^2)$, $B = 0.059(P/\mu a^2)$.
4. With $C = F_0 l^4/EI\pi^4$, $X = x/l$, $y = C(-\pi X + \frac{1}{2}\pi^3 X^2 - \frac{1}{6}\pi^3 X^3 + \sin \pi X)$.
5. Particles travel along straight lines with constant velocity.
7. The equations to be solved, together with suitable boundary conditions are the metric together with

$$\varphi \equiv 0, \qquad r\theta' = A, \qquad (R^2 - r^2)^{\frac{1}{2}} r' = B.$$

8. $E_1 = \frac{1}{4}E_0$.
9. $E_0 = \frac{1}{2}\omega\hbar$.
10. β is a root of

$$\alpha\beta(2\beta + 3) = 2(2\beta + 1)^3 \text{ with } \alpha = 8mV_0 a^2/\pi^2.$$

Chapter 10

1. $7 + 3 + 8 + 2 + 4 + 2 + 0 + 6 + 8 = 40$.
2. Two routes at a cost of 8 units.
3. Two policies each with profit 22, *KKRKKK*, *KRKRKK*.
4. In successive batches produce 4, 1, 1, 2, 2 at a total cost of 36 units.
5. With mesh size $\delta x = \delta y = \frac{1}{4}$ the minimum cost is 0.569 compared with the exact value of 0.642.
6. For (10.12) the recurrence relation becomes

$$C_k(x_k) = \min_{u_k} \{f(t_k, x_k, \dot{x}_k, u_k) + C_{k+1}[x_k + \Delta t g(t_k, x_k, u_k)]\}.$$

7. With $\Delta t = 0.1$ instead of (10.16)

$$x_{n+1} = 0.818\,7x_n + 0.090\,6u_n;$$

C_9 and u^*_9 remain as before, $u^*_8 = -0.680\,7x_8$ and $C_8(x_8) = 16.19x_8{}^2$.

Chapter 11

2. $u \equiv 0$, $x = e^{at}$.
3. k satisfies $k^4 = -1$. The boundary conditions are easy to fit but the resulting 4×4 matrix would need a numerical inversion. For the final time $0 = 2T + 1 + \lambda(T)x_2(T)$.
4. $u = 0.082\,6 \sinh (14)^{\frac{1}{2}}(t - 1)$,
   $x = -0.016\,5 \sinh (14)^{\frac{1}{2}}(t - 1) + 0.030\,9 \cosh (14)^{\frac{1}{2}}(t - 1)$.
7. $u = 0.372t - 1.068$, $x = 0.062t^3 - 0.534t^2$, $y = 0.186t^2 - 1.068t$.

9. Switching function

$$S(t) = \begin{cases} -\dfrac{C}{\beta_{max}} \log\left(\dfrac{m}{m_s}\right), & \beta = \beta_{max} \\ \dfrac{c}{m_s}(t_s - t), & \beta = 0 \end{cases}$$

where m_s is the mass at the switch time t_s.

Chapter 12

1. With $y = A(1 - x^2)$ method (i) gives $A = 0.579$ and method (ii) $A = 0.753$.
2. Each case leads to a formidable numerical problem. The best that can be done is to compare the work required in each method. The only problem that is feasible by hand is the Galerkin method on (i); with $u = x + Ax(1 - x)$, A takes possible values $-0.042\,75$, -121.5.
3. (i) $3u_0{}^2\alpha^4 + 4(\cosh u_0 - 1)\alpha^2 - K(2u_0) = 0$ with K defined in section 8.6.
   (ii) $\alpha^2 = 2(\cosh u_0 - 1)/u_0{}^2$.
   For small u_0 these give respectively α^2 as $\frac{1}{3}$, $\frac{1}{4}$, to order u_0.
4. With $v = A - \frac{1}{4}(x^2 + y^2)(P/\mu)$ least squares gives $A = \frac{1}{3}(Pa^2/\mu)$. (Compare the exact value $v(0, 0) = (Pa^2/\mu)0.295$.) The Trefftz method is unable to cope with the constant A and this must be fitted at some point or by the least squares method. Using $v = A - \frac{1}{4}(P/\mu)(x^2 + y^2) + B(x^4 - 6x^2y^2 + y^4)$ gives $B = -0.072\,9(P/\mu a^2)$ and fitting A gives $A = (Pa^2/\mu)0.275$.
5. Using the functions 1 and $r \sin \theta$ both methods give
   $T = \frac{1}{2}T_0 + (2T_0/\pi a)r \sin \theta$.
6. With $y = A[1 - (x/a)]$ the two methods give eigenvalue approximations of 2.74 and 2.53.
7. An approximation of the form $u = A + Bt + Ct^2$ gives the exact solution $A = 4/\pi$, $B = C = 0$.
8. (i) f satisfies the Bessel equation $r^2f'' + rf' + (r^2\lambda^2 - 1)f = 0$.
   (ii) No solution of this form exists, $u \sim r$ for small r. Try $u = r(a - r)g(\theta)$ to give $g'' + (\frac{1}{5}\lambda^2a^2 - 2)g = 0$.

References

AKHIEZER, N. I. (1962). *The calculus of variations*, Blaisdell Publishing Co, New York.

AOKI, M. (1971). *Introduction to optimization techniques*, Macmillan, New York.

BELLMAN, R. E. and DREYFUS, S. E. (1962). *Applied dynamic programming*, Princeton University Press.

BEVERIDGE, G. S. G. and SCHECHTER, R. S. (1970). *Optimization: theory and practice*, McGraw-Hill, New York.

BOURNE, D. E. and KENDALL, P. C. (1967). *Vector analysis*, Oldbourne, London.

BOX, M. J. (1966). *The Computer Journal*, **9**, 67.

BROYDEN, C. G. (1967). *Mathematics of Computation*, **21**, 368.

CARROLL, C. W. (1961). *Operations Research*, **9**, 169.

COURANT, R. and HILBERT, D. (1953). *Methods of mathematical physics*, vol. 1, Interscience, New York.

DANZIG, G. B. (1963). *Linear programming and its extensions*, Princeton University Press.

DAVIDON, W. C. (1959). AEC Research and Development Report ANL-5990.

FIACCO, A. V. and MCCORMICK, G. P. (1964). *Management Science*, **10**, 360.

FIACCO, A. V. and MCCORMICK, G. P. (1966). *Management Science*, **12**, 816.

FINLAYSON, B. A. and SCRIVEN, L. E. (1966). *Applied Mechanics Reviews*, **19**, 735.

FLETCHER, R. (1968). ICI Ltd, Management Services Report MSDH/68/19.

FLETCHER, R. (1970). *The Computer Journal*, **13**, 317.

FLETCHER, R. (1972). See Lootsma (1972).

FLETCHER, R. and POWELL, M. J. D. (1963). *The Computer Journal*, **6**, 163.

GELFAND, I. M. and FOMIN, S. V. (1963). *Calculus of variations*, Prentice-Hall, Englewood Cliffs, New Jersey.

HASTINGS, N. A. J. (1973). *Dynamic programming with management applications*, Butterworth, London.

HIMMELBLAU, D. M. (1972). See Lootsma (1972).

HUANG, H. Y. (1970). *Journal of Optimization Theory and Applications*, **5,** 405.

KANTOROVICH, L. V. and KRYLOV, V. I. (1958). *Approximate methods in higher analysis*, Noordhoff, Amsterdam.

KEMBLE, E. C. (1958). *The fundamental principles of quantum mechanics with elementary applications*, Dover, New York.

KUHN, H. W. and TUCKER, A. W. (1951). Non-linear programming, J. Neyman (Editor), in *Proceedings of 2nd Berkeley symposium on mathematical statistics and probability*, University of California Press, Berkeley, California.

LOOTSMA, F. A. (Editor) (1972). *Numerical methods for non-linear optimization*, Academic Press, London.

LUENBERGER, D. G. (1969). *Optimization by vector space methods*, Wiley, New York.

LUSH, P. E. and CHERRY, T. M. (1956). *Quarterly Journal of Mechanics and Applied Mathematics*, **9,** 6.

MIRSKY, L. (1955, 1963). *An introduction to linear algebra*, Oxford University Press, 1955, 2nd edition, 1963.

MURRAY, W. (Editor) (1972). *Numerical methods for unconstrained optimization*, Academic Press, London.

NELDER, J. A. and MEAD, R. (1965). *The Computer Journal*, **7,** 308.

NOTON, A. R. M. (1965). *Introduction to variational methods in control engineering*, Pergamon Press, Oxford.

PARKINSON, J. M. and HUTCHINSON, D. (1972). See Lootsma (1972).

POWELL, M. J. D. (1970). *Society for Industrial and Applied Mathematics Review*, **12,** 79.

PROTTER, M. H. and WEINBERGER, H. F. (1967). *Maximum principles in differential equations*, Prentice-Hall, Englewood Cliffs, New Jersey.

ROSENBROCK, H. H. (1960). *The Computer Journal*, **3,** 175.

SAGE, A. P. (1968). *Optimum systems control*, Prentice-Hall, Englewood Cliffs, New Jersey.

SCHECHTER, R. S. (1967). *The variational method in engineering*, McGraw-Hill, New York.

SPENDLEY, W., HEXT, G. R. and HIMSWORTH, F. R. (1962). *Technometrics*, **4,** 441.

WALSH, J. (1971). *Proceedings of the Royal Society, Ser. A*, **323,** 155.

WILDE, D. J. (1964). *Optimum seeking methods*, Prentice-Hall, Englewood Cliffs, New Jersey.

ZIENKIEWICZ, O. C. and CHEUNG, Y. K. (1967). *The finite element method in structural and continuum mechanics*, McGraw-Hill, New York.

Index